T0206049

Demand Forecasting for Executives and Professionals

This book surveys what executives who make decisions based on forecasts and professionals responsible for forecasts should know about forecasting. It discusses how individuals and firms should think about forecasting and guidelines for good practices. The book introduces readers to the subject of time series, presents basic and advanced forecasting models, from exponential smoothing across ARIMA to modern Machine Learning methods, and examines human judgment's role in interpreting numbers and identifying forecasting errors and how it should be integrated into organizations.

This is a great book to start learning about forecasting if you are new to the area of forecasting or have some preliminary exposure to it. Whether you are a practitioner, either in a role managing a forecasting team or operationally involved in demand planning, a software designer, a student, or an academic teaching business analytics, operational research, or operations management courses, the book can inspire you to re-think demand forecasting.

No prior knowledge of higher mathematics, statistics, operations research, or forecasting is assumed in this book. It is designed to serve as a first introduction to the non-expert who needs to be familiar with the broad outlines of forecasting without specializing in it. This may include a manager overseeing a forecasting group, or a student enrolled in an MBA program, an executive education course, or programs not specializing in analytics. Worked examples accompany the key formulae to show how they can be implemented.

Key Features:

- While there are many books about forecasting techniques, very few are published targeting managers. This book fills that gap.
- It provides the right balance between explaining the importance of demand forecasting and providing enough information to allow a busy manager to read a book and learn something that can be directly used in practice.
- It provides key takeaways that will help managers to make a difference in their companies.

Dr. Stephan Kolassa is a Data Science Expert at SAP, Switzerland and Honorary Researcher at Lancaster University, UK.

Dr. Bahman Rostami-Tabar is an Associate Professor in Data and Management Science, at Cardiff University, UK.

Prof. Enno Siemsen is the Patrick A. Thiele Distinguished Chair in Business, University of Wisconsin-Madison, USA.

Demand Forecasting for Executives and Professionals

Stephan Kolassa, Bahman Rostami-Tabar
and Enno Siemsen

CRC Press is an imprint of the
Taylor & Francis Group, an informa business
A CHAPMAN & HALL BOOK

Designed cover image: © Marjolein Poortvliet

First edition published 2024
by CRC Press
6000 Broken Sound Parkway NW, Suite 300, Boca Raton, FL 33487-2742

and by CRC Press
4 Park Square, Milton Park, Abingdon, Oxon, OX14 4RN

CRC Press is an imprint of Taylor & Francis Group, LLC

Library of Congress Cataloging-in-Publication Data

Names: Kolassa, Stephan (Data science expert), author. | Rostami-Tabar, Bahman, author. | Siemsen, Enno, author.
Title: Demand forecasting for executives and professionals / Stephan Kolassa, Bahman Rostami-Tabar and Enno Siemsen.
Description: First edition. | Boca Raton, FL : CRC Press, 2024. | Includes bibliographical references and index.
Identifiers: LCCN 2023013054 (print) | LCCN 2023013055 (ebook) | ISBN 9781032507736 (hardback) | ISBN 9781032507729 (paperback) | ISBN 9781003399599 (ebook)
Subjects: LCSH: Economic forecasting. | Decision making.
Classification: LCC HB3730 .K615 2024 (print) | LCC HB3730 (ebook) | DDC 330.01/12--dc23/eng/20230703
LC record available at https://lccn.loc.gov/2023013054
LC ebook record available at https://lccn.loc.gov/2023013055

ISBN: 978-1-032-50773-6 (hbk)
ISBN: 978-1-032-50772-9 (pbk)
ISBN: 978-1-003-39959-9 (ebk)

DOI: 10.1201/9781003399599

Typeset in Latin Modern font
by KnowledgeWorks Global Ltd.

Publisher's note: This book has been prepared from camera-ready copy provided by the authors.

Contents

Preface

Forecasting is the art and science of predicting the future, not with a magical crystal ball, but with proven statistical methods based on objective data. With more data and an improved understanding of forecasting methods, decisions become more influenced by forecasts.

We can forecast many things, from tomorrow's weather to financial markets to election results. This book focuses on one specific use case for forecasting: forecasting demand for products or services.

There are many books and articles to learn from (see Chapter 22). This book is not an in-depth technical treatise. Instead, it aims to bring you up to speed with minimal pain, prepare you to learn more, or have intelligent conversations about forecasting with experts. Read the Preface to find out whether this book is for you.

This book is also available online at https://dfep.netlify.app/, where we will update it continually.

Happy forecasting!

Who this book is for

This book is a high-level introduction to demand forecasting. It will, by itself, not turn you into a forecaster. It gives you an overview of the most common forecasting methods and the mindset behind forecasting. This book is for you if you

- Are a manager responsible for forecasters
- Are an IT professional whose responsibilities include administering a forecasting system
- Are an executive making decisions based on the forecasts generated by someone else in your work
- Contemplate learning more about forecasting without delving deep into the details (yet)
- Want to learn about forecasting and need a companion book to give you an overview, along with more technical in-depth materials

How to read this book

We divide this book into six parts:

- The "Introduction" gives an overview, setting the scene and explaining the basic philosophy of this book: forecasting is indispensable for making decisions under uncertainty.
- "Forecasting basics" discusses the essential toolbox of forecasters: knowing your time series, time series features, and time series decomposition into seasonal and trend components.
- "Forecasting models" gives a very high-level introduction to standard models: simple methods, Exponential Smoothing, ARIMA, causal models, count data and intermittent demand, and Artificial Intelligence/Machine Learning.
- "Forecasting quality" explains how to assess the quality of a forecast (which holds some pitfalls for the unwary) and what forecasting competitions are.
- "Forecasting organization" discusses organizational and human aspects of forecasting: what to look for in a forecaster, how to deal with forecasters, how to build a forecasting team, and why forecasting fails.
- Finally, "Learning more" points to various resources to learn more, from textbooks at various levels of technical detail, across organizations and events to websites.

While we did our best to provide good information and advice on forecasting, your forecasts are your own, so use them responsibly. If you create and act on a forecast, we disclaim any responsibility for any adverse outcomes. This book is a learning guide; it will not automatically make your forecasts as accurate as you would like them to be.

Acknowledgments

We want to thank Business Expert Press, publishers of our earlier book (Kolassa and Siemsen 2016), for allowing us to reuse much of the material in this new book and, more generally, for being great publishers. Special thanks to Scott Isenberg for his guidance over the years.

In addition, we would like to thank CRC Press for allowing us to publish an online version of this book.

We thank Marjolein Poortvliet, who created a very inspirational title page for this book and endured our requests for changes and adaptations.

We are also grateful to the people who created and maintained the free tools we used in preparing this book, from the statistical programming environment R (R Core Team 2022) and the packages we used (Xie 2022; Hyndman and Kourentzes 2018; Wickham et al. 2019; Schauberger and Walker 2022; Zhu 2021; Arnold 2021; Hyndman et al. 2021; Hyndman 2020; Stoffer and Poison 2023; Slowikowski 2023) to LaTeX (The LaTeX community, n.d.).

Thanks are also very much due to Nicolas Vandeput, who pointed out a number of shortcomings in a draft of this book.

SK would like to express his gratitude to his long-suffering but patient family, Iris, Sophie and Philipp, who did not see much of him when this book was on the home stretch. He is also thankful to his co-authors for a long, sometimes grueling, but never boring journey together.

BRT would like to thank his partner, Maryam, for her love and continuous patience during the whole process of writing this book. He also expresses his gratitude to his co-authors for their valuable contributions and the opportunity to collaborate on this book.

ES deeply thanks his wife, Min, and children, Oskar and Thalia. Their love has always kept him going. ES also thanks his co-authors, his team, and Dean Samba for allowing him to work on this project despite managing a portfolio of graduate programs.

Finally, we are grateful to you, our reader, for investing your time and energy in reading our book. We hope you will learn a lot and even occasionally will be entertained!

Introduction

1

Introduction

Why do we forecast at all? Forecasts are necessary for any decision under uncertainty. Good forecasts can be very valuable. Bad forecasts can threaten your company's very survival. This chapter discusses the value of a good forecasting process and briefly examines forecasting software.

1.1 The value of a good forecasting process

It is common to become frustrated about forecasting. The necessary data is often dispersed throughout the organization. The algorithms used to analyze this data are often opaque. Those within the organization trained to understand the algorithms often do not understand the business, and those who breathe the business do not understand the algorithms. The actual forecast is then discussed in long and unproductive consensus meetings between diverse stakeholders with often conflicting incentives; in between, the forecast is often confused with goals, targets, and plans. The resulting consensus can be a political compromise far removed from any optimal use of information. Decision-makers often ignore these forecasts and instead come up with their own *guess* since they do not trust the forecast and the process that created it. Even if the forecasting process works well, the inherent demand uncertainty often creates actuals far from the forecast. It is hard to maintain clarity in such a setting and not become frustrated by how challenging it is to rely on forecasts.

Yet, what alternative do we have? Eliminating the forecasting process within an organization will only create worse parallel shadow processes. Every plan, after all, needs a forecast, whether that forecast is an actual number based on facts or just the gut feeling of a planner. Some companies can change their business model to a make-to-order system, eliminating the need to forecast demand and manufacture their products to stock. However, this alternative model still requires ordering components and raw materials based on a forecast, planning capacity, and training the workforce according to an estimate of future demand. Even a make-to-order system relies on an *implicit* forecast, namely that demand will stay sufficiently stable for the system to keep working.

A central metric for every supply chain is how long it would take for all partners in the supply chain to move one unit – from the beginning to the end – into the market. This metric shows the total lead time in the supply chain. If customers are not willing to wait that long for a product, a supply chain cannot change to a complete make-to-order system. Someone in the supply chain will need to forecast and hold inventory. If that forecasting system fails, the supply chain will feel the resulting costs and disruptions.

Every manager involved in forecasting must accept that there are no good or bad forecasts. There are only good or bad ways of creating or using forecasts. Forecasts should contain all the relevant information available to the organization and its supply chain about the market. Information is everything that reduces uncertainty. If a forecast is far from the actual demand, but the process that generated the forecast used all available information, the organization had bad luck. Conversely, if a forecast is spot on, but the process that created it neglected important information, the organization was lucky but should consider improving its forecasting process. In this sense, bad forecasts can only result from bad forecasting processes. As with decision-making under uncertainty generally, one should not question the quality of the decision or forecast itself given the actual outcome; one should only examine the process that led to this decision or forecast. Betting money in roulette on the number 20 does not become a bad choice just because the ball rolls onto a different number – and neither does it become a better choice if the ball happens to land on the 20!

Different time series are more or less predictable. If a series has much unexplainable variation, there is a limit to how well we can forecast it. Figure 1.1 offers an example of two time series that are very different in their forecastability. While a good forecasting process will lead to better predictions for a time series by explaining some variation in the series, there are limits to the inherent predictability of such series. Repeated inaccurate forecasts may be a sign of a bad forecasting process but may also result from excessive noise in the underlying demand. Accordingly, a good forecasting process is not necessarily one that predicts a time series perfectly but one that improves the forecasts for a series compared to simple forecasting methods (which can be pretty competitive indeed, see Chapter 8). We will discuss the concept of forecastability in more detail in Chapter 5 and examine how to deal with failure in the forecasting process in Chapter 21.

Many organizations rely exclusively on point forecasts. A point forecast is a single number – an estimate of what an unknown quantity will most likely be. It is, however, improbable that the actual number will be exactly equal to the point forecast. Thus, one always needs to consider and deal with the remaining uncertainty in the forecast. Ideally, one should conceptualize a forecast as a probability distribution. That distribution can have a center, usually equal to the point forecast. Yet that distribution also has a spread, representing the

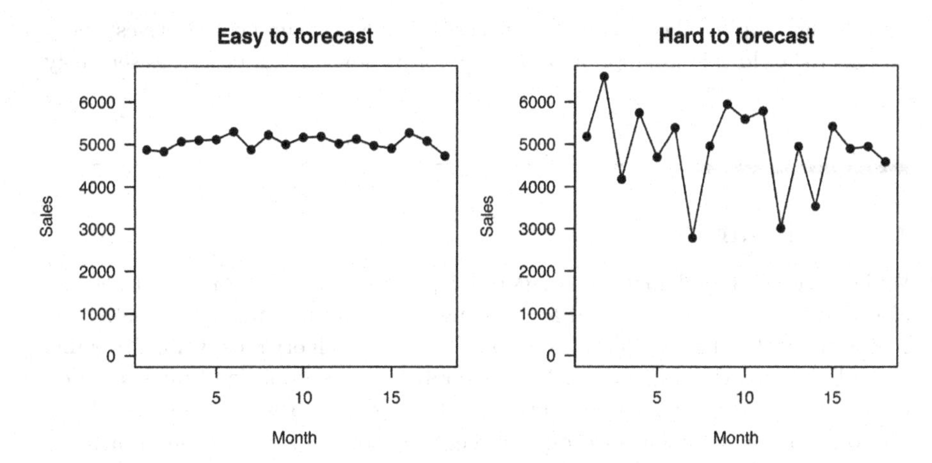

FIGURE 1.1
Easy and hard to forecast time series

remaining uncertainty around the point forecast. Good forecasting processes will communicate this spread effectively; bad forecasting processes will remain silent on this issue, projecting unrealistic confidence in a single number. Further, failing to communicate the inherent forecast uncertainty can lead to decision-makers using highly uncertain and highly certain forecasts similarly. It is common, for example, for firms to require equal safety stocks across different products, even though the uncertainty inherent in these products may vary vastly. The root cause of this problem often lies in insufficient reporting of uncertainty. We will further explore the idea of probabilistic forecasting in Chapters 3 and 4.

The effective design of forecasting processes seems complicated, but the benefits of getting the forecasting process right are tremendous. Fixing the forecasting process is a managerial challenge that usually does not require significant financial investments. The challenge of improving the forecasting process often does not lie in investing in advanced machines or information technology or in the costs of hiring more people and expanding the organization. Instead, the challenge is managing cross-functional communication and pushing through change despite many stakeholders (J. Smith 2009).

Yet, the returns can be huge if an organization can overcome these challenges. For example, Oliva and Watson (2009) documented that the improvement of a forecasting process at an electronics manufacturing company led to a doubling of inventory turns and a decrease of 50% in on-hand inventory. Similarly, Clarke (2006) documents how the major overhaul of the forecasting process at Coca-Cola Inc. led to a 25% reduction in days of inventory. These are supply chain improvements that would otherwise require significant investments in technology; if an organizational change (though challenging and time-consuming) of the forecasting process can achieve similar objectives, every manager should take the opportunity to improve forecasting processes seriously.

1.2 Software

While we often highlight the managerial aspects of forecasting in this book, we also delve into the statistics of forecasting. Our goal in doing so is to provide managers with a basic intuition on how forecasting algorithms work – to shine some light into this black box. In this context, we emphasize that this book does not assume the use of any particular forecasting software. There is a large set to choose from when selecting a forecasting software, and a comprehensive review of the features, strengths, and weaknesses of all commercially available products is beyond the scope of this book. For an overview, interested readers may visit the OR/MS biannual survey of forecasting software; the most recent one was authored by Schaer et al. (2022).

Especially in the earlier chapters of this book, we will refer to functions in Microsoft Excel to help readers implement some ideas. This spreadsheet modeling software is widely available, and most managers will have a copy installed on their laptops or tablets. However, Excel suffers from inaccuracies in its statistical and optimization functions (Mélard 2014). We would not wager that Microsoft has addressed these issues in the more recent version of Excel. Further, the standard functionality of Excel only allows for very limited time series analysis and forecasting. Therefore, using Excel for forecasting inevitably requires some coding and manual entry of formulas. Maintaining a consistent forecasting process in Excel is challenging. Spreadsheets start accumulating errors and undocumented changes over time (Singh 2013). When implemented correctly, spreadsheets have the advantage of being very transparent. Commercially available forecasting software, on the contrary, can often have a black-box character. Excel is a good complementary tool for forecasting – learning, communicating, and testing out new ideas – but it should not become a standard, permanent tool for forecasting in an organization.

An important alternative is the free statistical computing environment R (https://www.r-project.org/, R Core Team 2022). While R is more difficult to learn and use than Excel, its functionality is much broader. Through user-written content, many existing forecasting methods are free in R (Kolassa and Hyndman 2010). Furthermore, Integrated Developing Environments (IDE) like RStudio (https://posit.co/) make the software more accessible, and excellent introductory books to R from a forecasting perspective are available (see Section 22.2). The `forecast` (Hyndman et al. 2023) and `fable` packages (O'Hara-Wild et al. 2020) are the gold standard of forecasting.

Similarly, many Data Scientists nowadays use Python, a ubiquitous programming language. One key advantage of Python over R is that software or IT professionals outside the forecasting world are much more likely to be familiar with it, making it easier to share code across functional boundaries. Python does have some time series forecasting functionalities. However, its forecasting strengths lie more in modern Machine Learning tools (see Chapter 14), and not so much in the classic time series tools. Nixtla (https://github.com/Nixtla) produced a lot of time series tools that are fast and perform well compared to various alternatives. Packages like `statsforecast` (Federico Garza 2022), `pmdarima` (T. G. Smith et al. 2022), `sktime` (Löning et al. 2019) mainly aim at replicating (and therefore invariably lag behind) functionalities from R.

Key takeaways

1. Almost every business decision is about the future and is thus based on forecasts. We need forecasts. We cannot eliminate forecasts, but we can question whether we have an effective forecasting process that uses all available information within our organization and supply chain.

2. Different time series will differ in how hard they are to predict. Inaccurate forecasts may result from an ineffective forecasting process or may be due to the unpredictable nature of a particular business.

3. No forecast is perfect. We need to directly confront, quantify, and manage the uncertainty surrounding our forecasts. Failure to communicate this uncertainty makes risk management associated with the forecast ineffective.

4. Fixing the forecasting process can lead to considerable improvements in the organization and supply chain without significant technological investments. The challenge is to manage cross-functional communications and overcome organizational silos and conflicting incentives.

2

The forecasting workflow

Producing forecasts to inform decisions is an iterative process involving several potentially repeated steps. As shown in Figure 2.1, the forecasting workflow starts by identifying the decision that the forecast supports. Once we understand what decision the forecast informs, we need to derive the requirements we need to meet. We next gather data and information, prepare and visualize data, choose and train models, and produce the forecast. We can evaluate its performance and iterate to improve. Or we stop the iteration, communicate the forecast to stakeholders and possibly adjust the forecast by incorporating human judgment. This chapter outlines these steps in detail.

2.1 Identify a decision requiring a forecast

Why prepare a forecast? If asked this question, most managers usually say, "to inform our decisions." A forecast has no purpose per se. At least one, if not many, decisions within the organization require the forecast. This purpose determines the forecasting requirements. Any forecasting task should therefore start with identifying operational (short-term), tactical (mid-term), and strategic (long-term) decisions which require forecasts.

Examples of strategic decision and planning processes include network design, capacity planning, distribution channels, and workforce planning. At the tactical level, example decisions include product/service assortments, staffing, inventory optimization, and sales & operations planning (S&OP). At the operational level, firms make decisions related to production and workforce scheduling, distribution, transportation, and delivery plans. Forecasters must understand how the forecast is used, who requires it, and how it fits in with the organization. This initial step is essential and only sometimes straightforward.

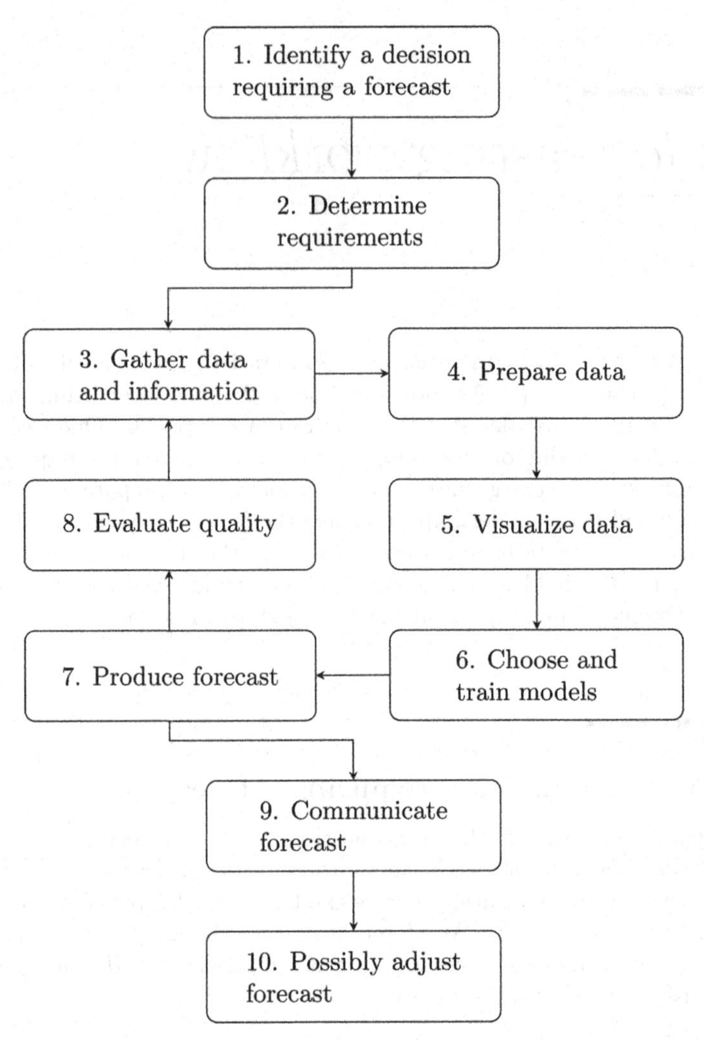

FIGURE 2.1
The forecasting workflow

2.2 Determine requirements

Each decision identified in the first step may require forecasting different data at different granularities, horizons, frequencies, and structures. Decisions will directly dictate the forecasting requirements, which should at least include the following:

TABLE 2.1
Planning and decision processes and forecast requirements

Level	Decision	Forecast horizon	Temporal granularity	Cross-sectional granularity	Forecast type
Strategic	Make-or-buy	Years	Yearly	Total	Point/interval
Strategic	Enter new market	Years	Yearly	Total	Point/interval
Tactical	Assortment plan	Months/years	Monthly	Category	Point
Tactical	Promotion plan	Weeks/months	Weekly	Brand	Point
Operational	Retail store reordering	Days/weeks	Daily	SKU	Quantile
Operational	Staffing plan	Days/weeks	Hourly/daily	Per worker type	Point

- The *forecast variable* is what we intend to forecast and is also known as the *response*, *output*, or *dependent variable*. This variable should not be confused with the decision that the forecast intends to inform.
- The *forecast horizon* is how far into the future we forecast.
- The *forecast frequency* defines how often we generate the forecast.
- The *temporal granularity* is the time granularity we forecast (e.g., daily, weekly, and monthly totals).
- The *cross-sectional granularity* at which we forecast (e.g., SKU or product group, store or region; see Chapter 13, especially Section 13.1).

Table 2.1 gives examples of the attributes of our forecasts driven by the decision.

In addition to determining the forecast requirements, we should consider some additional elements at this stage. For instance: how much can the forecast cost? How should we measure and report forecast accuracy? Does the forecast improvement add value to the business? Is the forecasting method easy to understand and interpretable? Should we use specific forecasting support systems and software? We cover these topics in different sections of this book.

2.3 Gather data and information

Once we have determined the forecasting requirements, we need to gather the relevant data and information that we will use to develop our forecasting model. Various types of data can be helpful at this stage:

1. *Past/historical time series data on the variable we intend to forecast.* If we want to forecast the sales volume of a product in the future, we need to collect past sales volume. Many organizations record historical data at finer time granularity or arbitrary timestamps. Forecasting usually requires some temporal aggregation. We discuss this in detail in Section 13.3.

2. *Past and future data about deterministic predictors. Predictors* are data that help us forecast a time series (see Chapter 11). A predictor is *deterministic* if its future values are known or determined in advance. Examples include different types of promotions or prices, public holidays, festivals, or sports events. Deterministic predictors can also be master data, which do not change over time, such as colors or sizes of products.

3. *Past and future data about stochastic predictors.* A predictor is *stochastic* if we cannot know its future values precisely. For instance, while we know the past weather, future weather is unknown. Therefore, we must forecast the future values of stochastic predictors if we include them in a forecasting model. Suppose our forecast depends on lagged values of a stochastic predictor (see Section 11.3). For example, tomorrow's headache medication sales depend on today's sports game outcome. In that case, we do not need to forecast the predictor.

4. *Expertise of individuals in an organization and any contextual information that may affect the forecast variable.* Domain knowledge of those who collect data, analyze it, produce forecasts, or use them to inform decisions is a valuable source of information. Such knowledge will help you tell the story behind the data, incorporate it into models and check the validity of the information. It can also be the reason why we adjust the output of our statistical forecasting. Sometimes, we may be unable to obtain enough data to use our models. Without data, we rely on judgmental forecasting methods alone (see Chapter 16), which incorporate domain knowledge of people in the organization and up-to-date information about future events.

2.4 Prepare data

Once we have the necessary raw data, we need to ensure it is a reliable source of information and is in the correct format for analysis and forecasting. Forecasting models may have different data requirements; some require the series to be in time order with no gaps, and others require no missing values. Checking our data is essential to understanding its features and should always be done before we train our models on data. Often, raw data has many problems, such as formatting issues, errors, or missing values. Data preparation generally involves importing data into the software, tidying the data, and checking and fixing errors, missing values, duplicated observations, and temporal gaps.

Raw data can be messy in infinite ways, and technical data science skills are often insufficient to prepare data for analysis. Using the knowledge of those who collect or work with data is crucial in preparing data. See Section 19.1 on what skills this requires of the forecaster. Section 5.3 contains a relevant discussion about data quality.

2.5 Visualize data

Before developing our model for a given dataset, we should conduct an initial analysis by plotting our data using data visualization tools. Plotting data will allow us to identify essential features such as trends or seasonality. If the data has no systematic features, producing accurate forecasts is challenging. We can use time series graphs to check data quality and identify unusual observations, such as extreme values and outliers. If something unusual occurs in the data, try to figure out what happened. Is it a product launch? Is it a special event (e.g., Black Friday)? Or is it an error in the data? Visualizing data helps to highlight potential issues and then take appropriate actions to fix them before choosing and training models. We introduce various time series graphs in Chapter 6 to help us visualize data.

2.6 Choose and train models

Which forecasting model should we choose? The answer depends on several factors such as the type of decision the forecast intends to inform, forecasting requirements, data availability, the strength of the relationships between predictors and the forecast variable, how important explainability is to us, and so on. It is common to choose, train and compare the performance of multiple models. Each model may have different assumptions and account for different types of systematic information. Instead of picking just one, we might use multiple models and combine their forecasts (see Section 8.6).

Once we have chosen the model to use in forecasting, we need to train it on the historical data. Training is the estimation of parameters using historical data so that the model can forecast the output as close as possible to the actual value. We examine some forecasting models in Chapter 8 to Chapter 16.

2.7 Produce forecast

Once we have selected and trained our models, we can use them to produce forecasts for the required horizon. We can represent the forecast as a point forecast, a prediction interval, or a probability density (see Chapter 4). If the model uses only the historical time series to forecast, we must provide the number of future observations we want to forecast. If our model uses additional information, such as predictors, we need to provide a dataset of future values of those predictors to generate the forecast. We discuss such causal models in Chapter 11.

2.8 Evaluate quality

After training a model on the historical data and producing forecasts, it is critical to investigate its performance. In this step, we want to ensure our model has not failed in capturing the systematic information and did not overfit by following unpredictable random errors. We can evaluate the forecast accuracy of the model after observing actual data for the forecast period. We provide more detail on assessing a model's accuracy in Chapter 17.

However, "quality" also has dimensions besides accuracy, like runtime, storage requirements, data requirements, or explainability. A method may provide superior forecasting accuracy but run very slowly or be completely opaque except to experts. Such properties would hinder its acceptance among business users.

2.9 Communicate forecast

Two factors are essential to increase acceptance and use of forecasting in practice: (a) *what* forecasts we communicate to stakeholders and (b) *how* we communicate them. Traditionally, we represent forecasts as point forecasts. Such a representation does not acknowledge uncertainty about the future or give the information required for a decision-maker to manage risk. Communicating forecasts as distributions or prediction intervals is often preferable (See Chapter 4).

For effective communication, forecasters need to know how to display forecasts. Visualization is critical when communicating forecast distributions so stakeholders can appreciate the information. Moreover, communicating the forecast variable and its impact on decisions is vital. For instance, providing the forecast distribution of patient admissions using a forecasting model is essential in an Accident and Emergency department. Still, such a distribution per se is insufficient for decision-making. Complementing this information with the cost, number of staff required, and waiting times resulting from the demand forecasts and potential actions is more valuable (see Section 17.5).

2.10 Possibly adjust forecast

A forecast adjustment indicates an expert's judgment about the final forecast. Adjusting statistical demand forecasts using people's expertise is very common. Forecast adjustments could be beneficial when we know that the forecasting model omits some important information or new information has become available. In those cases, we can make adjustments. We should avoid minor adjustments and focus only on the important ones – and document the reasons for any adjustments. Over time, we can improve our models by incorporating key drivers to deliver a better baseline forecast in the first place. We devote Chapter 16 to the role of judgment in forecasting.

Key takeaways

1. The forecasting workflow should always start with identifying decisions that require forecasts.
2. Forecasting requirements must be clearly defined. We need to ensure that forecasts are being generated at the right frequency, over appropriate horizons, at the right level of temporal (e.g., at hourly,

daily, weekly, etc.) and cross-sectional aggregation (e.g., at product group or SKU level).

3. When building a model, we start with the historical time series data, identify potential deterministic or stochastic predictors and their lags, and any qualitative information available across the organization.
4. Data preparation for time series analysis and forecasting can include checking and fixing errors, missing values, duplicated observations, and temporal gaps.
5. Communicating forecasts in a format that acknowledges uncertainty and accuracy implications on decisions (i.e., business value) is critical.

3

Choice under uncertainty

Forecasts are indispensable tools to make decisions under uncertainty. There are various different forecasts a method can output. In increasing order of sophistication, these are point forecasts (i.e., single number forecasts), prediction intervals and predictive distributions. Knowing when to use what is crucial, especially if we apply the forecast to ensure target service levels.

3.1 Forecasting methods

A forecast is an input to support decision-making under uncertainty. Forecasts are created by a statistical model or by human judgment. A statistical model is an algorithm, often embedded into a spreadsheet or other software, that converts data into a forecast. Of course, choosing which algorithm an organization uses for forecasting and how the organization implements this algorithm often requires the use of human judgment as well. But in the present context, we use the term "judgment" to indicate that humans influence or override a statistical forecast (see Chapter 16). Human judgment is the intuition and cognition decision-makers can employ to convert all available data and tacit information into a forecast. We discuss statistical models and human judgment in much more detail later in this book.

In reality, most forecasting processes contain elements of both judgment and statistics. A statistical forecast may serve as the basis of discussion, but this forecast is then revised in some form or other through human judgment to arrive at a consensus forecast, a combination of different existing forecasts (statistical or judgmental) within the organization. A survey of professional forecasters found that while 16% of forecasters relied exclusively on human judgment and 29% depended solely on statistical methods, the remaining 55% used either a combination of judgment and statistical forecast or a judgmentally adjusted statistical forecast (Fildes and Petropoulos 2015). Another study of a major pharmaceutical company found that more than 50% of the forecasting experts answering the survey did not rely on the company's statistical models when preparing their forecast (Boulaksil and Franses 2009). Forecasting in practice is thus not an automated process. People usually influence the process. However, often the sheer number of forecasts means that not all can be inspected by humans. For instance, retail organizations may need to forecast more than 20,000 SKUs daily for hundreds or thousands of stores to drive replenishment. Naturally, forecasting at this level tends to be a more automated task, and determining *which* forecasts need judgmental input becomes key.

Throughout this book, we refer to a *forecasting method* as the process through which an organization generates a forecast. This forecast does not need to be the final consensus forecast, although some method always generates the consensus forecast. The beauty of any forecasting method is that its accuracy can be judged ex-post and compared against other methods. There is always an objective realization of demand that we can (and should) compare to the forecast to create a picture of forecast accuracy over time. Of course, we should never base this comparison on small samples. However, suppose a forecasting method repeatedly forecasts far from the actual demand realizations. In that case, this observation is evidence that the method is not working well, particularly if there is evidence that another method works better. In other words, whether a forecasting method is good or bad is not a question of belief but of scientific empirical comparison. If the forecastability of the underlying demand is challenging, then multiple methods will fail to improve accuracy. We will explore how to make such a forecasting comparison in more detail in Chapter 18.

One key distinction to keep in mind is that the forecast is not a target, a budget, a plan, or a decision. A forecast is simply an expression or belief about the likely state of the future. Targets, budgets, or plans are decisions based on forecasts, but these concepts should not be confused with the forecast itself. For example, our point forecast for demand for a particular item we want to sell on the market may be 100,000 units. However, it may make sense for us to set our sales representatives a target of selling 110,000 units to motivate them to do their best. In addition, it may also make sense to plan to order 120,000 units from our contract manufacturer since there is a chance that demand is higher. We want to balance the risk of stocking out against having excess inventory. To make the latter decision effectively, we would include data on ordering and sales costs, as well as assumptions on how customers would react to stockouts and how quickly the product may or may not become obsolete.

Single-number forecasts: point forecasts

Most firms operate with *point forecasts* – single numbers that express the most likely outcome on the market. Yet, we all understand that such a notion is somewhat ridiculous; the point forecast is unlikely to be realized precisely. A forecast can have immense uncertainty. Reporting only a point forecast communicates an illusion of certainty. Let us recall a famous quote by Goethe: “To be uncertain is to be uncomfortable, but to be certain is to be ridiculous.”

Prediction intervals

Point forecasts, in this sense, are misleading. It is much more valuable and complete to report forecasts in the form of a probability distribution or at least in the form of *prediction intervals*, that is, a "best case" and a "worst case" scenario – always with the understanding that reality may still fall outside these bounds, if only with a very small probability. The endpoints of a prediction interval are also known as *quantile forecasts*. For instance, a 10% quantile forecast is a number such that we expect a 10% chance that the future observation falls below it. A 90% quantile forecast is an analogous number, which will of course be higher than our 10% quantile forecast. And if we combine a 10% quantile forecast and a 90% quantile forecast, we have an 80% (= 90% − 10%) prediction interval.

Creating such prediction intervals requires a measure of uncertainty in the forecast. While we will explore this topic in detail in Chapter 4, we provide a brief and stylized introduction here. We usually express uncertainty as a standard deviation (usually abbreviated by the Greek letter σ). Given a history of forecast errors, measuring this uncertainty is pretty straightforward. The simplest form would be to calculate the population standard deviation (using the `=STDEV.P` function in Microsoft Excel) of past observed forecast errors (see Chapter 17 for a more in-depth treatment of measuring the accuracy of forecasts). Assuming that forecast errors are roughly symmetric, that is, over-forecasting is as likely and extensive as under-forecasting, we can then conceptualize the point forecast as the center (i.e., the mean, median, and most likely value, abbreviated by the Greek letter μ) of a probability distribution, with the standard deviation σ measuring the spread of that probability distribution.

We illustrate these concepts in Figure 3.1. Suppose our forecasting method generated a point forecast of 500. We calculate a standard deviation of our past forecast errors as 20. We can thus conceptualize our forecast as a probability distribution with a mean of 500 and a standard deviation of 20 (in this illustration, we assumed a normal distribution, drawn with the Excel function =NORM.DIST). A probability distribution is nothing but a function that maps possible outcomes to probabilities.

For example, one could ask what the probability is that demand is between 490 and 510. The area under the distribution curve between the x values of 490 and 510 would provide the answer – or the Excel function call `=NORM.DIST(510, 500, 20, TRUE)-NORM.DIST(490, 500, 20, TRUE)`. The result is about 38.3%: the probability of seeing demand between 490 and 510.

Given a forecasted mean and a standard deviation (and the crucial assumption of a normal distribution), we can thus calculate a 95% prediction interval, by calculating the 97.5% quantile of the distribution as the upper limit (or

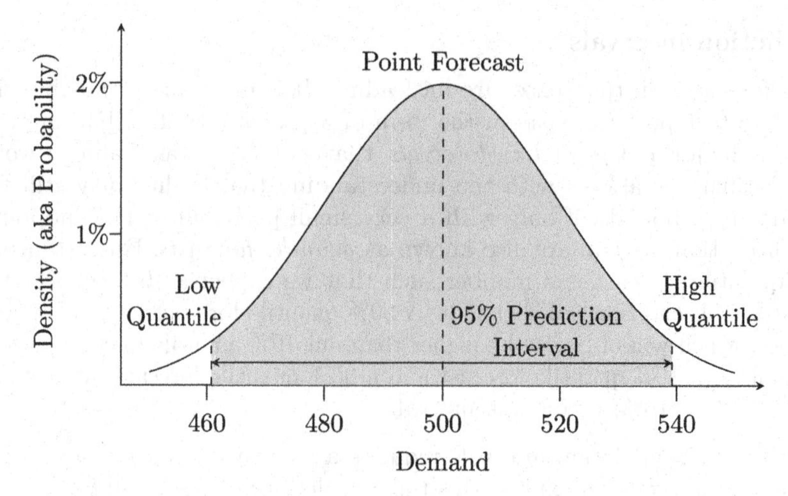

FIGURE 3.1
Forecasts from a probabilistic perspective

quantile) and the 2.5% quantile as the lower limit, which yields a theoretical coverage of $97.5\% - 2.5\% = 95\%$. That is, if our assumptions are correct, we predict that 95% of future observations fall within this interval. In Excel, we can calculate these interval limits or *quantile forecasts* using `=NORM.INV(0.025, 500, 20)` and `=NORM.INV(0.975, 500, 20)`, respectively, yielding an interval of $(460.8, 539.2)$. We can examine the effects of various changes to the input variables. For instance, increasing the standard deviation makes our assumed time series harder to forecast (compare Figure 1.1) and will make the prediction interval wider, as will increasing the target interval coverage above 95%. We immediately see how targeting a higher service level than 97.5% (the upper limit of our interval) requires a higher safety stock.

A caveat: calculated prediction intervals are often too narrow (Chatfield 2001). One reason is that these intervals often do not include the uncertainty of choosing the correct model and the uncertainty of the environment changing. It was surprising in the recent M5 forecasting competition that prediction intervals actually had the required coverage (Makridakis et al. 2022)!

Our prediction interval has non-integer limits. While product is sold in integer units, the upper quantile forecast is 539.2. This inconsistency is a result of using the normal distribution. This distribution is commonly used in statistics and forecasting because it makes the mathematics *much* simpler, and because it is usually sufficient as an approximation. Using this distribution starts causing problems if we have very slow selling items, so-called *count data*. See Chapter 12 for a discussion of this situation.

Note that symmetry of forecast errors is not always the case in practice. For example, items with low demand are naturally censored at zero, creating a skewed error distribution (see Chapter 12); similarly, if political influence within the organization creates an incentive to over- or under-forecast, the distribution of errors can be skewed. Chapter 17 will examine how to detect such forecast bias in more depth.

Finally, a word about terminology: prediction intervals are often confused with confidence intervals (Soyer and Hogarth 2012). While a confidence interval represents the uncertainty about an estimate of a parameter of a distribution (i.e., the mean of the above distribution), a prediction interval represents the uncertainty about a realization from that distribution (i.e., demand taking specific values in the above distribution). The confidence interval for the mean is much smaller (depending on how many observations we used to estimate the mean) than the prediction interval characterizing the distribution. In forecasting, we almost always look at prediction intervals, not confidence intervals.

Predictive distributions

A predicted probability distribution (known either as a *predictive distribution*, a *predictive density* or a *distributional forecast*) thus communicates a good sense of the possible outcomes and the associated uncertainty with a forecast.

Ideally, decisions that use forecasts can work with full predictive distributions. However, how should we *report* such a probabilistic forecast? Usually, we do not draw the actual distribution since this may be too much information to digest for decision-makers. Reporting a standard deviation in addition to the point forecast can already be challenging to interpret. A good practice is to report 95 or 80% prediction intervals as caclulated above, that is, intervals in which we are 95 or 80% sure that demand will fall.

Prediction intervals thus provide a natural instrument for forecasters to adequately communicate the uncertainty in their forecasts and for decision-makers to decide how to manage the risk inherent in their decision. Reporting and understanding prediction intervals and predictive distributions requires some effort. However, we can automate the calculation and reporting of these intervals in modern forecasting software. Not reporting (or ignoring) the uncertainty in forecasts can have distinct disadvantages for organizational decision-making. Reading just a point forecast without a measure of uncertainty gives you no idea how much uncertainty there is in the forecast. The consequences of not making forecast uncertainty explicit can be dramatic. At best, decision-makers will judge how much uncertainty is inherent in the forecast. Since human judgment in this context generally suffers from over-confidence (Mannes and Moore 2013), not making forecast uncertainty explicit will likely lead to under-estimating the inherent forecast uncertainty, leading to less-than-optimal safety stocks and buffers in resulting decisions. At worst, decision-makers will treat the point forecast as a deterministic number and ignore the inherent uncertainty in the forecast entirely, as well as any precautions they should take in their decisions to manage their demand risk.

Adopting prediction intervals in practice is challenging; one criticism often brought up is that it is difficult to report more than one number and that the wide range of a 95% prediction interval makes the interval almost meaningless. This line of reasoning, however, misinterprets the prediction interval. The range also contains information about probabilities since values in the center of the range are more likely than those at the end. We can easily visualize these differences, e.g., in a *fan plot* as in Figure 3.2, which shows a demand time series from the M5 competition (Makridakis, Spiliotis, and Assimakopoulos 2022): areas of different shading indicate different pointwise prediction intervals. It is common to plot an inner 80% prediction band and an outer 95% prediction band as in this figure, i.e., areas in which we believe 80% respectively 95% of future observations will fall. Fan plots thus provide an easily interpreted summary of our forecast's uncertainty and represent the state-of-the-art in communicating forecast uncertainty (Kreye et al. 2012).

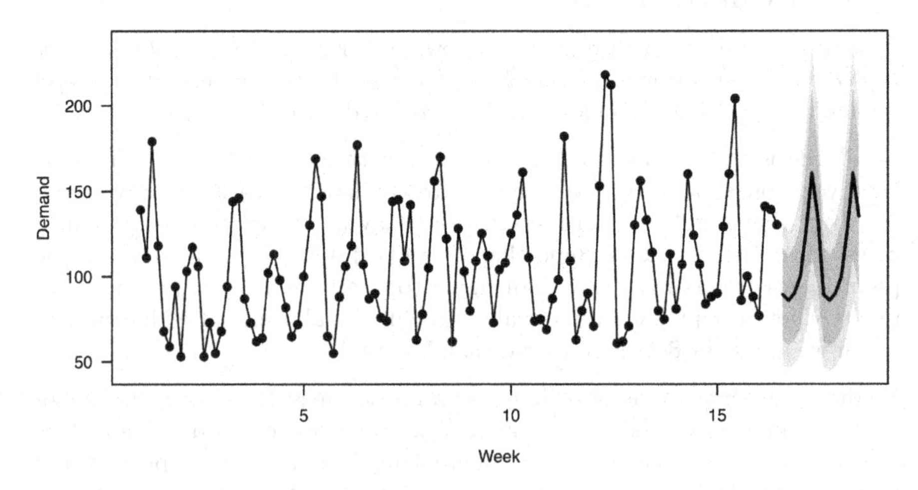

FIGURE 3.2
Daily demand for one SKU at one store, with point forecasts and a fan plot

Another argument against using prediction intervals is that, in the end, we need single numbers for decision-making. Ultimately, containers must be loaded with a particular volume; capacity levels require hiring a certain number of people or buying a certain number of machines. What good is a forecast showing a range when we need one number in the end? This argument makes the mistake of confusing the forecast with decision-making. The forecast is an input into a decision, not a decision per se. Firms can and must set service-level targets and solve the inherent risk trade-offs to translate probabilistic or interval forecasts into single numbers and decisions.

3.2 Forecasts and service levels

Consider the following illustrative example. Suppose you are a baker and need to decide how many bagels to bake in the morning for selling throughout the day. The variable cost to make bagels is 10 cents, and you sell them for $1.50. You donate bagels you do not sell during the day to a kitchen for the homeless. You forecast that demand for bagels for the day has a mean of 500 and a standard deviation of 80, giving you a 95% prediction interval of roughly (340, 660). How many bagels should you bake in the morning? Your point forecast is 500 – but you probably realize this would not be the correct number. Baking 500 bagels would give you just a 50% chance of meeting all demand during the day.

This 50% chance represents an essential concept in this decision context – the so-called type I service level or in-stock probability, that is, the likelihood of meeting all demands with your inventory (see textbooks on inventory control, such as Silver, Pyke, and Thomas 2017, on different notions of service levels). This chance of not encountering a stockout is a crucial metric often used in organizational decision-making. What service level should you strive to obtain? The answer to that question requires carefully comparing the implications of running out of stock with the repercussions of having leftover inventory – that is, managing the inherent demand risk. The key factors here are so-called *overage* and *underage* cost, that is, assessing what happens when too much or too little inventory is available.

An overage situation in the case of our bagel baker implies that they have made bagels at the cost of 10 cents that they are giving away for free; this would lead to a loss of 10 cents. An underage situation implies that they have not made enough bagels and thus lose a profit margin of $1.40 per bagel not sold. This underage cost is an opportunity cost – a loss of $1.40. Assuming no other expenses of a stockout (i.e., loss of cross-sales, loss of goodwill, loss of reputation, etc.), this $1.40 represents the total underage cost in this situation. Underage costs, in this case ($1.40), are much higher than overage costs (10 cents), implying that you would probably bake more than 500 bagels in the morning.

Finding the right service level in this context is known as the *newsvendor problem* in the academic literature (Qin et al. 2011). Its solution is simple and elegant. One calculates the so-called critical fractile, the ratio of underage to the sum of underage and overage costs. In our case, this critical fractile roughly equals 93% (=1.40/[1.40+0.10]). The critical fractile is the optimal type I service level. In other words, considering the underage and overage costs, you should strive for a 93% service level in bagel baking. In the long run, this service level balances the opportunity costs of running out of stock with the obsolescence costs of baking bagels that do not sell.

So how many bagels should you bake? We have seen above how to calculate this in Microsoft Excel: the formula `=NORM.INV(0.93, 500, 80)` gives us a result of about 618. In other words, you should bake 618 bagels to have a 93% chance of meeting all demand in a day, which is the optimal target service level.

This example illustrates the difference between the forecast, which serves as an input into a decision, and the actual decision, which is the number of bagels to bake. The point forecast is not the decision, and making a good decision would be impossible without understanding the uncertainty inherent in the point forecast. Good decision-making under uncertainty requires understanding uncertainty and balancing risks. In addition, while forecasts are crucial ingredients for decision-making, there are always additional inputs, usually in the form of costs and revenues. In the case of our bagel baker, the critical managerial task was not to influence the forecast but to understand the cost factors involved with different risks in the decision and then define a service level that balances these risk factors. A good forecast in the form of a probability distribution or a prediction interval should make this task easier. The actual quantity of bagels baked is simply a function of the forecast and the target service level derived from the cost structure.

Our discussion highlights the importance of setting adequate service levels for the items in question. Decision-makers should understand how their organization derives these service levels. Since the optimal service level depends on the product's profit margin, items with different profit margins require different service levels. While overage costs are comparatively easy to measure (cost of warehousing, depreciation, insurance, etc.; see Timme and Williams-Timme 2003), underage costs involve understanding customer behavior and are thus more challenging to quantify. What happens when a customer desires a product that is not available? In the best case, the customer finds and buys a substitute product, which may even be sold at a higher margin, or puts the item on backorder. In the worst case, customers take their business elsewhere and tweet about the bad service experience. Studies of a mail-order-catalog business show that the indirect costs of a stockout – the opportunity costs from lost cross-sales and reduced long-term sales of the customer – are almost twice as high as the lost revenue from the stockout itself (Anderson, Fitzsimons, and Simester 2006). Similar consequences of stockouts threaten the profitability of supermarkets (Corsten and Gruen 2004). Setting service levels thus requires studying customer behavior and understanding the revenue risks associated with a stockout. Since this challenge can appear daunting, managers can sometimes react in a knee-jerk fashion and set very high service levels, e.g., 99.99%. Such an approach leads to excessive inventory levels and related inventory-holding costs. Achieving the right balance between customer service and inventory holding requires a careful analysis and thorough understanding of the business.

Note that while we use inventory management as an example of decision-making under uncertainty, a similar rationale for differentiating between forecast and related decision-making also applies in other contexts. For example, in service staffing decisions, the forecast relates to customer demand in a specific time period. The decision is how much service capacity to put in place. Too much capacity means resources wasted in firms. Too little capacity means wait times and possibly lost sales due to customers avoiding the service system due to inconvenient queues. Similarly, in project management, a forecast may involve predicting how long the project will take and finding how much time buffer we need to build into the schedule. Too large a buffer may result in wasted resources and lost revenue, whereas too little buffer may lead to projects exceeding their deadlines and resulting in contractual fines. One must understand that forecasting by itself is not risk management – forecasting simply *supports* managers in making decisions by carefully balancing the risk inherent in their choice under uncertainty.

Existing inventory management systems are more complex than examples in this chapter may suggest, since they require adjusting for fixed costs of ordering (i.e., shipping and container filling), uncertain supply lead times (e.g., by ordering from overseas), best-by dates and obsolescence, optimization possibilities (like rebates or discounts on large orders), contractually agreed order quantities, as well as dependent demand items (i.e., scheduling production for one unit requires ordering the whole bill of materials). A thorough review of inventory management techniques is beyond the scope of this book, and we refer interested readers to Nahmias and Olsen (2015), Silver, Pyke, and Thomas (2017) or Syntetos et al. (2016) for comprehensive information.

Key takeaways

1. Forecasts are not targets, budgets, plans, or decisions; these are different concepts that need to be kept apart within organizations, or confusion will occur.

2. Often the word *forecast* is shorthand for *point forecast*. However, point forecasts are seldom perfectly accurate. We need to measure and communicate the uncertainty associated with our forecast. To accomplish this, we calculate prediction intervals or predictive densities, which we can visualize using fan plots.

3. A key concept to convert a forecast into a decision is the service level, that is, the likelihood of meeting an uncertain demand with a fixed quantity. Target service levels represent critical managerial decisions and should balance the risk of not having enough units available (underage) and having too many units (overage).

4

A simple example

To bring forecasting to life, our objective in this chapter is to provide readers with a simple example of applying and interpreting a statistical forecasting method in practice. This example is stylized and illustrative only. In reality, forecasting is messier, but this chapter serves as a guideline for thinking about and successfully executing and applying a forecasting method for decision-making.

4.1 A point forecast

Suppose you are interested in predicting weekly demand for a particular productor service. Specifically, we have simulated historical data from the past 50 weeks and want to get a forecast for how demand for your product will develop over the next 10 weeks, i.e., weeks 51 to 60 (see Figure 4.1). Completing this task means making ten different forecasts; the one-step-ahead forecast is your forecast for week 51, the two-step-ahead forecast is your forecast for week 52, and so forth. We will assume that we know that the data has no trend or seasonality (since we simulated it, we know this is true) to keep things simple. Trends and seasonality are predictable patterns in the data. We will examine them in more detail in Chapter 7. Further, no additional market information is available beyond this demand history (Chapter 11 will examine how to use additional information to achieve better predictions). We show a time series plot of our data in Figure 4.1.

Here are two simple approaches to create a point forecast (we cover simple forecasting methods in more detail in Chapter 8): either take the most recent observation of demand (= 3370) or calculate the long-run average over all available data points (= 2444). Both approaches ignore the distinct shape of the time series; note that the series starts low and then exhibits an upward shift. Calculating the long-run average ignores the observation that the time series currently hovers higher than in the past. Conversely, taking only the most recent demand observation as your forecast misses that the last observation in the series is close to an all-time high. Historically, an all-time high in the series usually does not signal a persistent upward shift but is followed by lower

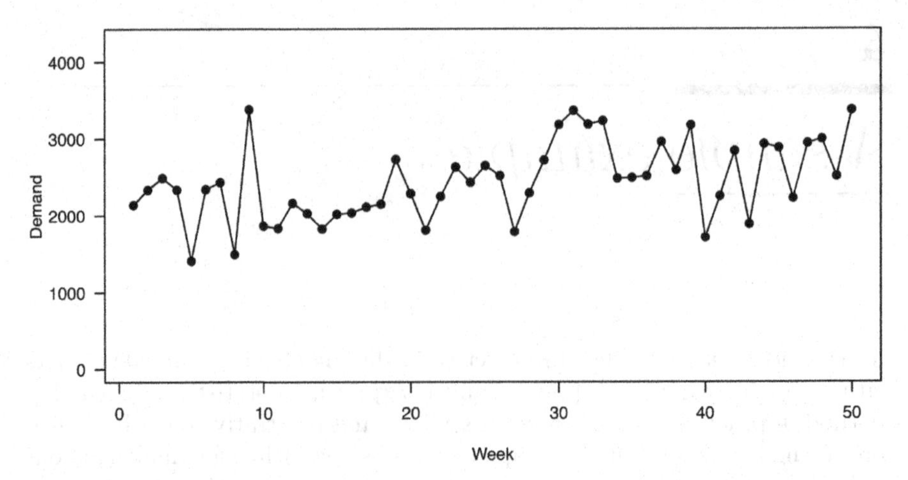

FIGURE 4.1
Time series for our example

demands in the following weeks. This sequence of events is a simple case of regression to the mean.

We could, of course, split the data and calculate the average demand over the last 15 periods (= 2652). Not a wrong approach – but the choice of how far back to go (= 15 periods) seems ad hoc. Why should the last 15 observations receive equal weights while those before receive no weight? Further, this approach does not fully capture the possibility that we could encounter a shift in the data; going further back into the past allows your forecast – particularly our prediction intervals – to reflect the uncertainty of further shifts in the level of the series occurring.

A weighted average over all available demand data would solve these issues, with more recent data receiving more weight than older data. What would be good weights? If I have 50 time periods of past data, do I need to specify 50 different weights? It turns out that there is a simple method to create such weighted averages, which also turns out to be the correct forecasting method to use for this kind of time series: (Single) Exponential Smoothing (SES). Chapter 9 will provide more details on this method; for now, we just need to understand that a vital aspect of this method is that it uses a so-called smoothing parameter α (alpha). The higher α, the more weight is given to recent as opposed to earlier data when calculating a forecast. The lower α, the more the forecast considers the whole demand history without heavily discounting older data. The more basic implementations of Exponential Smoothing require the user to specify α, but more sophisticated ones will optimize this parameter instead.

Suppose we feed the 50 time periods of this time series into one of these more sophisticated Exponential Smoothing tools to estimate an optimal smoothing parameter. In that case, we will obtain an α value of 0.15. This value indicates some degree of instability and change in the data (which we see in Figure 4.1).

Now you are probably thinking: Wait! I have to create a weighted average between two numbers, where one of these numbers is my most recent forecast. But I have not made any forecasts yet! If the method assumes a current forecast and I have not made forecasts in the past using this method, what do I do? The answer is simple. We use the most recent *model* or *in-sample fits*. In other words, suppose we had started in week 1 with a naive forecast of your most recent demand and applied the SES method ever since to create forecasts. What would have been your most recent forecast for week 50? Figure 4.2 depicts these in-sample fits.

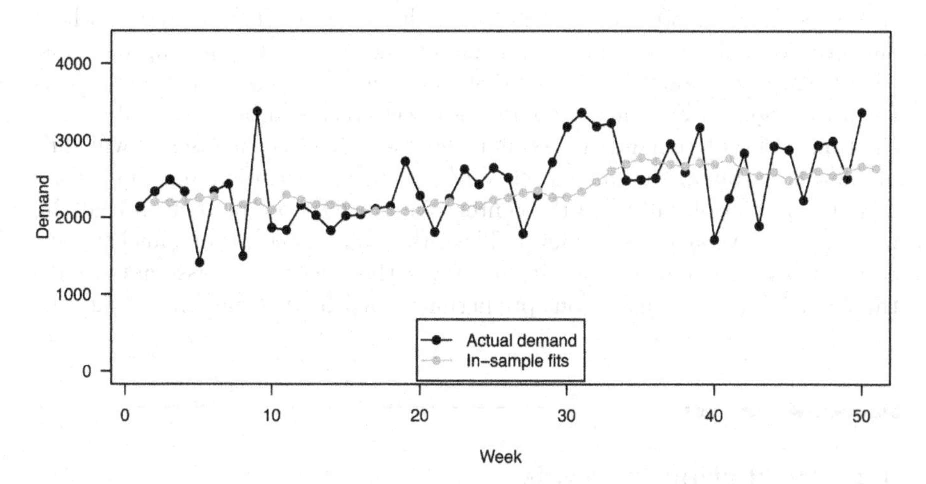

FIGURE 4.2
Model or in-sample fits for our example

These in-sample fits are not actual forecasts. First, we used the data from the last 50 weeks to estimate a smoothing parameter. Then, we used this parameter to generate fitted forecasts for the previous 50 periods. We would not have known this smoothing parameter in week 1 since we did not have enough data to estimate it. Thus, the recorded fitted forecast for week 2, supposedly calculated in week 1, would not have been possible to make in week 1. Nevertheless, this method of generating fitted forecasts now allows us to make a forecast for week 51. Specifically, we take our most recent demand observation (= 3370) and our most recent "fitted" forecast (= 2643) and calculate the weighted average of these two numbers ($0.15 \times 3370 + 0.85 \times 2643 = 2752$) as our point forecast for week 51.

So, what about the remaining nine point forecasts for periods 52 to 60? The answer is surprisingly simple. The SES model we are using here is a *level only* model. Such a model assumes that there is no trend or seasonality (which, as mentioned initially, is correct for this time series since we constructed it with neither trend nor seasonality). Without such additional time series components (see Chapter 7), the point forecast for the time series remains flat, i.e., the same number. Moving further out in the time horizon may influence the spread of our forecast probability distribution. Still, in this case, the center of that distribution and the point forecast remains the same. Our forecasts for the next 10 weeks are thus a flat line of 2752 for weeks 51 to 60. Our two-step-ahead, three-step-ahead, and later point forecasts equal our one-step-ahead point forecast.

Our intuition may tell us that this is odd; indeed, intuitively, many believe that the shape of the series of forecasts should be similar to the shape of the time series of demand (Harvey 1995). Yet this is one of those instances where our intuition fails us; the time series contains noise, which is the unpredictable component of demand. A good forecast filters out this noise (an aspect that is visible in Figure 4.2). Thus, the series of forecasts should be less variable than the time series of demand. We do not know whether the time series will shift up, down, or stay at the same level. Without such information, our most likely demand prediction falls into the center. We must keep a prediction that the time series stays at the same level. Thus, the point forecast remains constant as we predict further into the future. Only the uncertainty associated with this forecast may change as our predictions reach further into the unknown.

4.2 Prediction intervals

So how to calculate a prediction interval associated with our point forecasts? We will use this opportunity to explore three different methods of calculating prediction intervals:

1. using the standard deviation of observed forecast errors,
2. using the empirical distribution of forecast errors,
3. using theory-based formulas.

The first method is the simplest and most intuitive one. Calculating the errors associated with our past in-sample fits is easy by calculating the difference between actual demand and fitted forecasts in each period. We calculate the standard deviation of these forecast errors ($\sigma = 466.51$). This value represents the spread of possible outcomes around our point forecast (see Figure 3.1). If we want to calculate a prediction interval around the point forecast, we use this number in a calculation as given in Section 3.1. For instance, to obtain an 80% prediction interval in Microsoft Excel, we use the formulas `=NORM.INV(0.90, 2752, 466.51)` and `=NORM.INV(0.10, 2752, 466.51)`, for a result of (2154, 3350). In other words, we can be 80% sure that demand in week 51 falls between 2154 and 3350, with the most likely outcome being 2752.

Method (1) assumes that our forecast errors roughly follow a normal distribution; method (2) does not make this assumption but generally requires more data to be adequate. An 80% prediction interval ignores the top and the bottom 10% of errors we can make. Since we have approximately 50 "fitted" forecast errors, ignoring the top and the bottom 10% of our errors roughly equates to ignoring the top and the bottom five (=10% × 50) errors in our data. In our data, the errors within these boundaries are (–450; 582). Therefore, another simple way of creating an 80% prediction interval is to add and subtract these two extreme errors to and from the point forecast to generate a prediction interval of (2302, 3334), which is a bit more narrow than our previously calculated interval from method (1). In practice, bootstrapping techniques exist that increase the effectiveness of this method in terms of creating prediction intervals.

The final method (3) only exists for forecasting methods with an underlying statistical model – like the SES method we used in Section 4.1. For such methods, one can use the appropriate formula to calculate the standard deviation of the forecast error and derive a prediction interval. For example, in the case of SES, the procedure for this calculation is simple: If one predicts h periods into the future, the standard deviation of the forecast error for this prediction is the one-period-ahead forecast error standard deviation, as calculated in method (1), multiplied by $1 + (h - 1) \times \alpha^2$. Similar formulas exist for other methods but can be more complex; your forecasting software should be able to apply the correct calculations.

How does one generally calculate the standard deviation of forecast errors that are not one step ahead but h steps ahead? Suppose we prepare a forecast in week 50 for week 52 and use the standard deviation of forecast errors for one-step-ahead forecasts resulting from method (1). In that case, our estimated forecast uncertainty will generally under-estimate the actual uncertainty in this two-step-ahead forecast. Unless the time series is entirely stable and has no changing components, predicting further into the future implies a higher likelihood that the time series will move to a different level. The formula mentioned in method (3) would see the standard deviation of the one-step-ahead forecast error for period 51 (= 466.51) increase to 477.65 for the two-step-ahead forecast error for week 52 and 488.80 for the three-step-ahead forecast error for week 53. Prediction intervals increase in width accordingly.

Another approach would be calculating the two-step-ahead fitted forecasts for all past demands. In the case of Single Exponential Smoothing, the one-step-ahead forecast equals the two-step-ahead forecast. Still, our error calculation now requires comparing this forecast to demand one period later. The resulting error standard deviations for the two-step-ahead forecast (= 480.02) and three-step-ahead forecast (= 503.87) are higher than those for the one-step-ahead forecast and the adjusted standard deviations using the formula from method (3).

Forecasting software will usually provide one or more of these methods to calculate prediction intervals and will allow visualizing them, e.g., in a fan plot. For instance, Figure 4.3 shows point forecasts and 70% prediction intervals from method (3).

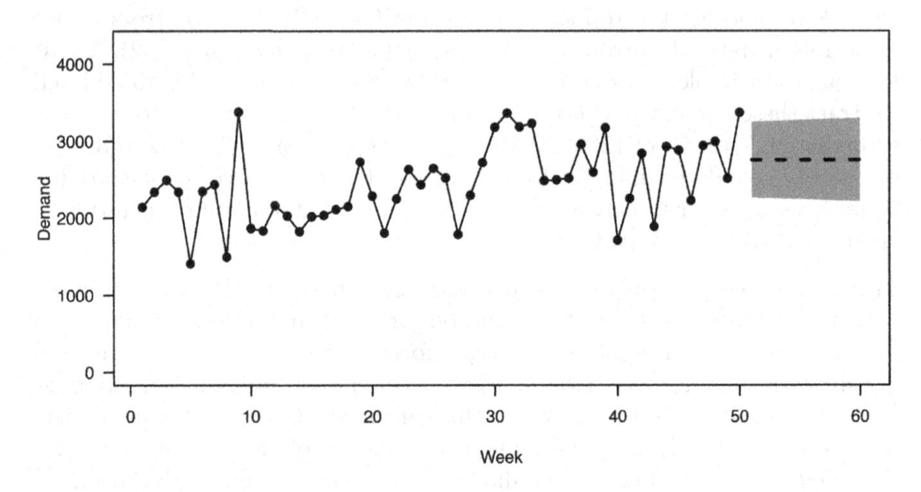

FIGURE 4.3
Demand, point forecasts, and prediction intervals

So what is the proper method to use? How do we come up with the correct prediction interval? Which standard deviation estimate works best? The common criticism against method (1) is that it under-estimates the uncertainty about the true forecasting model and does not incorporate the notion that the model could change. Standard errors are thus too low. Method (2) requires a lot of data to be effective. Method (3) is usually based on some assumptions and may not be robust if reality violates these assumptions. In practice, getting the right approach to calculate prediction intervals can require careful calibration and selection from different methods. Chapter 17 outlines how you can assess the accuracy of prediction intervals and compare different methods in practice. Generally, using any of the methods described here is better than using no method. While getting reasonable estimates of the underlying uncertainty of a forecast can be challenging, having any estimate is better than assuming that there is no uncertainty in the forecast or supposing that all forecasts have the same inherent uncertainty.

4.3 Predictive distributions

The preceding sections highlight that forecasts can consist of a single point (point forecasts) or an entire prediction interval. These different representations of a forecast are all stepping stones toward describing the probability density of future demand. We can go one step further and develop the complete probability distribution of each future outcome: the *predictive distribution.*

As a side note, prediction intervals are usually calculated based on an assumed density "under the hood." Specifically, the algorithm uses an assumption of a probability density, feeds in parameters (like the expectation or the standard deviation), and extracts the prediction interval endpoints. Figure 3.1 shows the relationship between a predictive density and the extracted prediction intervals.

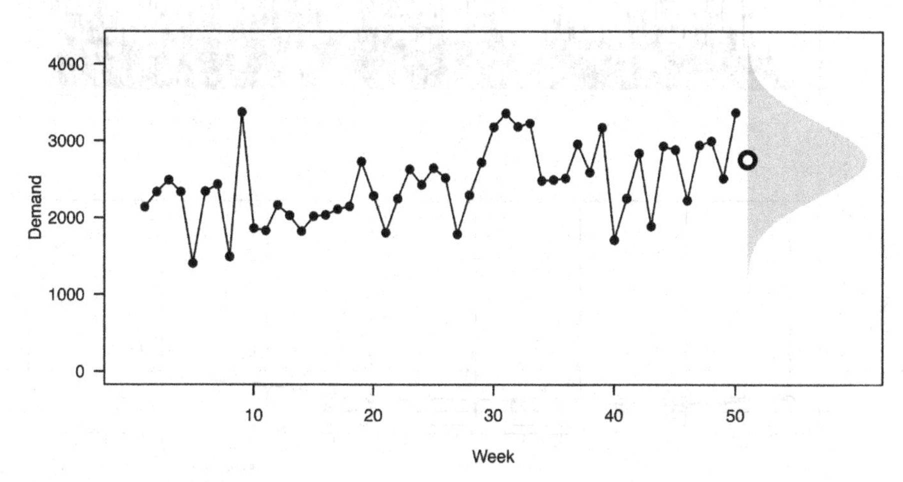

FIGURE 4.4
Demand, point forecast, and predictive density for the next observation

The most commonly assumed predictive density is a normal (or "Gaussian") distribution, i.e., the familiar bell curve, as shown in Figure 3.1. Figure 4.4 depicts our ongoing time series example, emphasizing the point forecast and the predictive density for the next observation. We rotated the bell curve by 90 degrees to align it with our time series plot. We could also calculate and plot point forecasts and predictive densities for periods further ahead in the series. Since the point forecast is the same for all future periods (compare Figure 4.3), all these densities would have the same center. But since we become less and less sure the further we predict into the future, the prediction intervals will get wider and wider, as we illustrated in Figure 4.3. Similarly, predictive densities that look farther ahead will also widen.

Realistically, predictive densities for different time points in the future can look very different. Figure 4.5 shows a time series of daily SKU demand at a single retail store from the M5 competition data. The time series exhibits day-of-week seasonality. More people tend to visit the store to purchase the product on weekends. We see in the plot that average sales tend to increase over the next week – but the predictive densities also widen. To satisfy this demand, we would need a higher safety stock on the weekend to achieve a target service level in the face of increasing predicted demand variance. A fixed safety stock for the week would lead to a lower service level (i.e., increased stockouts) on Fridays and Saturdays.

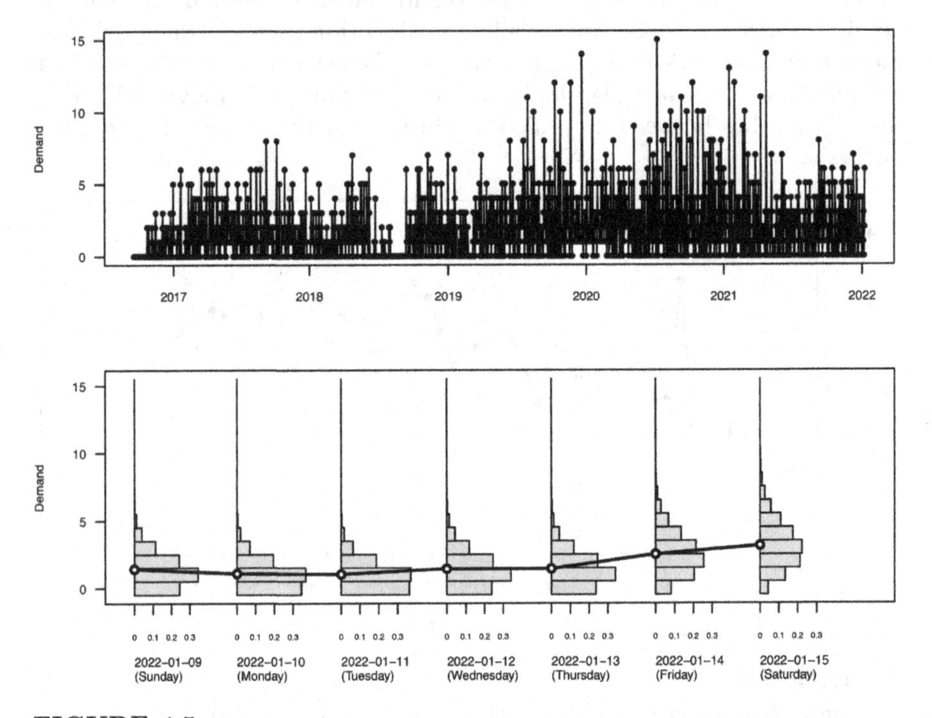

FIGURE 4.5
Daily SKU × store level retail sales: demand (top), point forecasts and discrete predictive densities (bottom) for the next seven observations

How can we calculate predictive distributions in practice? There are a number of commonly used methods. First, many standard time series forecasting methods will output a point forecast (as above) and an estimate of the standard deviation of future observations around that point forecast. If we can assume a normal (Gaussian) distribution of future observations, then we can just plug in the point forecast as the mean and this standard deviation to obtain the full predictive density. Second, we can collect the residuals between the historical observations and the in-sample fit. If we can assume that these residuals are representative for future forecast errors, then arranging them around the point forecast yields a predictive distribution. Third, one can fit use statistical or Machine Learning methods to forecast not the expectation of future observations, but many quantiles (using a *pinball loss* as a loss function), i.e., forecasts that aim at exceeding the future observation with a prespecified probability of 1%, 2%, ..., 99%. The collection of many such quantile forecasts (almost) yields a full predictive density. However, as we see from this description, calculating a full predictive density is a somewhat bigger challenge than getting a single point forecast.

4.4 Decision-making

Given that we now understand how to estimate the parameters of a probability distribution of future demand, how would we proceed with decision-making? Suppose we aim to place an order right now with a lead time of 2 weeks. Suppose we also place an order every week so that the order we place now has to cover the demand we face in week 53. Further, suppose we have perishable inventory, so we do not need to worry about existing inventory quantities or stocks left over from week 52. The inventory we have available to meet the demand in week 53 is equal to what we order right now. In this case, we would provide a demand forecast with the three-week-ahead forecast (= 2752) and the three-week-ahead standard deviation (= 488.80). It is essential to match the correct forecast (three-week-ahead) with the right forecast accuracy measure (three-week-ahead, and not one-week-ahead, in this case). Suppose we want to satisfy an 85% service level; the quantile forecast associated with an 85% service level is 3260 units; according to our forecast, this order quantity would have an 85% chance of meeting all demand.

This discussion concludes our simple example. Key learning points for readers are understanding the mechanics of creating point forecasts and prediction intervals for periods in the future and how these predictions can translate into decisions. We will now proceed to provide more detailed explanations of different forecasting methods.

Key takeaways

1. Calculating a forecast is not necessarily rocket science.
2. Forecasts come in three versions in increasing order of sophistication: point forecasts, prediction intervals, and predictive densities.
3. Correct predictive densities encode all the information we have about the future of a time series, but we can already improve decision-making using good point forecasts and prediction intervals.

Forecasting basics

5

Know your time series

The first step in any analysis is to understand the data. In forecasting, this usually means understanding a time series – or many time series. In this chapter, we consider basic properties to look out for in your time series data, from data quality to the amount of data and its forecastability. We also consider the more technical issue of stationarity.

5.1 Data availability

A time series is a sequence of similar measurements taken at regular intervals. Examples of time series include the annual profits of Uber, quarterly unemployment rates in France, monthly donations to UNICEF, weekly sales at a retailer like Walmart, the daily Google stock price, or hourly hospital admission in the University Hospital of Wales. Every time series has two components: a time index, i.e., a variable that allows ordering observations from past to present, and a measurement, i.e., a numerical variable showing the observation corresponding to each time index.

Time series analysis, our topic in Chapters 8 to 16, means examining the history of the time series to obtain information about the future. An inherent assumption in time series analysis is that the past of a series contains information about the future of the same series. Causal models (Chapter 11) and Artificial Intelligence/Machine Learning algorithms (Chapter 14) use the information contained in another data series to predict the future of a focal time series.

One essential aspect of time series forecasting is that time needs to be *bucketed* into periods. Many demand forecasts are prepared monthly ("how much product will our customers demand next month?"), and thus, the time series requires aggregating data into monthly buckets. Note that a month is not an entirely regular time interval since some months have more days than others. Still, this non-regularity is inconsequential enough to be ignored for most applications. Some operational forecasting requires weekly, daily, and sometimes even quarter-hourly time buckets, for example, in the case of retail store-level forecasting

(Fildes, Ma, and Kolassa 2022; Fildes, Kolassa, and Ma 2022) or call center traffic forecasting (Ibrahim et al. 2016). As we shall explain later in this chapter, this temporal dimension of aggregation implies different degrees of statistical aggregation, making forecasting more or less challenging. It also raises the question of temporal hierarchies (see Section 13.3). At what level should the organization forecast, and how does the organization aggregate or disaggregate to longer or shorter segments of time?

Another vital aspect to understand about our forecast is the availability of relevant historical data. Most methods discussed in the following chapters assume that some demand history for a time series exists. For example, suppose Exponential Smoothing is used (see Chapter 9), and the model includes seasonality. If our data are on quarterly granularity, we need at least nine past quarters' observations to fit the model. And if we have monthly data, then we need a history of at least 17 months (Hyndman and Kostenko 2007). These are *minimum* requirements; if the time series is very noisy, reliable parameter estimation requires much more data to be useful for forecasting. If too little data is available, the risk of detecting seasonality where none exists is much higher than the risk of failing to detect seasonality if it exists (see Section 11.7). So-called *shrinkage* or *regularization* methods are available to better deal with seasonality in such settings (Miller and Williams 2003), or more generally, in situations with too little data.

This data requirement may be excessive in many business contexts, especially in industries with frequent introductions of new products. If product lifecycles are short, one must carefully assess the novelty of changes to the product portfolio. Do they represent new product introductions, that is, the introduction of a novel good or service incomparable to any existing product in the firm's portfolio? Or do they represent a semi-new product introduction, a modified version of a product the firm has sold before (Tonetti 2006)? In the former case, the methods we will discuss here do not hold; forecasts will require modeling the product lifecycle, which requires good market research and extensive conjoint analysis to have a chance of being successful (Berry 2010). In the latter case, forecasting can proceed as we discuss here, as long as some existing data is representative of the semi-new product.

For example, suppose the semi-new product is a simple engineering change of a previous product version. In that case, the history of the earlier version of the product should contain information about the semi-new version. The earlier time series can thus initialize the forecasting method for the semi-new product. The time series level may change if the change is in packaging or style. Still, other components of the series, such as the trend, seasonality, and possibly even the uncertainty in demand, may remain constant. Thus, the estimates of these components from the past can be used for the new model as initial estimates, significantly reducing data needs.

Similarly, suppose the semi-new product is a new variant within an existing category. In that case, the current trend and seasonality estimates at the category level may apply to the new variant. In other words, intelligent top-down forecasting in a hierarchy (see Chapter 13) can allow forecasters to learn about these time series components by looking at the collection of similar variants within the same category.

The first step in time series analysis is understanding what data underlies the series. We frequently use *sales* rather than *demand* data as the input for demand forecasting. However, there is a difference between sales and demand, which comes into play during stockouts. If inventory runs out, customers may still demand a product so sales may be lower than the actual demand. In such a case, customers may turn to a competitor, delay their purchase, or buy a substitute product. In the latter case, the substitute's sales are higher than its raw demand. Thus, demand forecasts based on historical sales data may be systematically wrong if this effect is significant and not taken into account. Modern forecasting software can adjust sales data accordingly if stockout information is recorded. The mathematics of such adjustments are beyond the scope of this book. We refer interested readers to Nahmias (1994) for further details.

Adjusting sales to estimate demand requires clearly understanding whether data represents sales or demand. Demand can be challenging to observe in business-to-consumer contexts. If a product is not on the shelf, it is hard to tell whether a customer walking through the store demands it. In online retail, we can observe demand if the website does not show inventory availability to the customers before they place an item into their shopping basket. However, if this information is presented to customers *before* they click on "purchase," demand is again challenging to observe.

Demand is easier to observe in business-to-business settings since sellers usually record customer requests. In modern Enterprise Resource Planning (ERP) software, salespeople typically work with an *available-to-promise* number. Running out of *available-to-promise* means that the seller will not convert some customer requests into orders; if the salespeople do not record these requests, databases again only show sales, not demand.

5.2 How much data is necessary?

The amount of data we need to produce reliable forecasts depends on the context. As mentioned earlier, if your data has a trend or seasonality, we require more data history to estimate forecasting models successfully. In time series, the length of the time series should not be less than the number of parameters in the time series. For instance, the time series should have 24 months for a model to capture monthly seasonality (Chapter 6). Similarly, more history is needed to obtain reliable estimates if your data is noisy. How much data we need also depends on our forecast horizon. For example, forecasting 30 days out will require less data than forecasting 365 days out.

We may have a long time series, but consider throwing away old data, as it seems less useful. At what time does historical data become not worth the bother? Data processing may gradually take longer each day. Processes, products, and markets change over time, making older data apparently less relevant for current usage. Maybe you consider using a 5-year rolling period and discard the rest of your data. We would encourage you not to pursue this plan. In general, throwing away data is not a good practice. The more time series data you have, the better your model becomes. Having more data is especially important if you represent the forecast as a probability density distribution where the distribution's tails provide crucial information to deal with risk. Having a longer history makes your time series more representative – it will include all the ups and downs that your market has experienced and may yet experience again. More time series data means more information about the past.

Also, many forecasting models, such as Exponential Smoothing, can handle the underlying process, product, or market changes. If you use such a model, there are no issues with using a long history of a time series as an input to the model because the model will recognize and adjust to those changes. If you throw away old data, you risk not being able to capture longer-term trends. You may also fail to get reasonable estimates for special events like Christmas. Lastly, your model may under-estimate the uncertainty in your time series.

Another dimension to consider is the number of predictors (see Chapter 11) you use in your model. While using longer historical time series is usually useful, using more predictors is not necessarily beneficial. Spurious relationships may appear, and valuable information becomes less apparent (see Section 11.7). It would be best to have a handful of quality predictors that can describe the underlying variation rather than a grab bag of predictors with lower predictive quality.

5.3 Data quality

Good quality data is indispensable for any data science task. This truism holds in a time series forecasting context. There are many ways in which data can be wrong, corrupted, "bad," or simply inexistent. How critical it is to have quality data will depend on what time series process you are modeling. How to detect and address data quality issues depends on the context. Here are a few examples:

- Figure 4.5 shows daily sales of a specific SKU in a particular store. We notice a period of zero sales about one-third of the way through the series. Given that this product is otherwise relatively fast selling, we suspect that these zero sales were not caused by sudden low demand but by a stockout. The data are *censored.* The problem may be minor in this case. But other products may have longer stockouts, and these may have an impact on the forecasting model. Imagine an out-of-stock around Christmas – will our forecast for next year's Christmas be zero because the system thought that demand decreased *systematically* in this time?
- If some sensor or process collects our time series, we may not have any data for some periods because the sensor or process broke down. Our data preparation pipeline should catch and alert us to such problems. If we forget sanity checking, we could feed missing data as zero observations into our forecasting model.
- Retailers generally record their historical *sales* but sometimes fail to register their historical *promotions.* Without knowing when promotions occurred in the past, it is tough to forecast the impact of future promotions.
- Data may be corrupted, especially if any processing steps are involved. If we pull our time series out of a database via manual selection, joining, and aggregation steps, it is easy for errors to creep in and remain undetected.

One frequently overlooked aspect of data quality is accuracy measurement. If we assess our forecast accuracy using a holdout sample, and the data in that holdout period is problematic, we may draw completely wrong conclusions. Imagine that we had a stockout and censored retail data during the holdout period – any forecast close to zero would look very good, despite actual unconstrained demand being likely much higher. Optimizing our forecasts against bad-quality data can thus lead to systematically biased predictions. Evaluating forecasts against stockout periods is an excellent recipe to produce forecasting models that predict zero demand – i.e., models that drive us out of business.

How do we systematically detect such data quality issues? As there are so many different ways in which data quality can deteriorate, we can only provide a few general hints here.

It is always a good idea to plot a time series – ideally on multiple levels of aggregation because sometimes data quality problems only become apparent in the aggregate (just as they are sometimes only visible at a finer granularity). Plot several series, get a cup of coffee, and flip through the series. If anything looks wrong, consider whether it is and what may cause it. And do the same for any holdout data you may have to ensure you are not training and evaluating your forecasting models on bad data. It is always enlightening to look particularly at those time series that a method forecasts exceptionally well or poorly. A single outlier due to a data entry error can significantly impact forecast accuracy. Therefore, such an error can also be easily found and corrected by filtering time series for bad forecasts (which may actually be good ones and only look bad because we evaluate them against problematic data).

Once we have identified data quality issues, what can we do about them? Again, this depends on the domain in which we are and the forecasting method. If we have found a problem with our data preparation step, we might be able to correct the preparation and re-run it, ending up with correct and clean data. However, there is not always such a happy ending to our data cleansing efforts. What do we do with problematic data when we can't pull the correct and clean data from somewhere?

Some methods, like regression models (see Chapter 11) or tree-based methods (see Section 14.2), have no problem with missing data, so we can simply remove the problematic data points. Other methods, like ARIMA (see Chapter 10) or Exponential Smoothing (see Chapter 9), have significant issues with missing data, so we can't just *remove* problematic data points. We must *replace* them. Unfortunately, what to replace such problematic data with can be a science all its own. We could take the average of the last valid data point before the problem and the first sound data point after the problem. However, if we have a long string of problems (e.g., if our sensor was out for a more extended period), that will give us a long period of imputed constant values, which can dominate our forecasting model. Alternatively, we can fit some model to the valid data and predict the problematic periods. In this case, we are back to square one: to replace our problematic data to fit a forecasting model, we must choose a good "data replacement model." But to decide whether such a model is "good," we need to rely on the sound data we do have.

All this can get complicated, and there is an art to knowing how much time and effort to invest in cleaning data (and searching for data problems in the first place) and subsequently modeling the cleaned data. The key is to remember that data cleansing is *always* necessary. It *always* takes longer than anticipated, but it can yield significant benefits. Often, cleaning the input data leads to more substantial forecast improvements than trying more complicated algorithms on less clean data.

5.4 Stationarity

One key attribute of a time series is *stationarity*. Stationarity indicates that the mean of demand is constant over time, that the variance of demand remains constant, and that the correlation between current and most recent demand observations (and other parameters of the demand distribution) remains constant. Stationarity, in essence, requires that the time series has stable properties when looked at over time.

Many time series violate these criteria; for example, a time series with a trend (see Chapter 7) is not stationary since the mean demand is persistently increasing or decreasing. Similarly, a simple random walk (see Chapter 7 and Section 8.2) is not stationary since mean demand randomly increases or decreases in every period. In essence, non-stationary series imply that demand conditions for a product change over time, whereas stationary series imply that demand conditions are very stable. Some forecasting methods work well only if the underlying time series is stationary. Others, such as the ARIMA methods discussed in Chapter 10, preprocess the data (in the case of ARIMA models, by differencing; see Section 10.2) in an attempt to make it stationary. Figure 5.1 provides a few examples to illustrate the difference between stationary and non-stationary time series.

We often transform time series to become stationary before they are analyzed. Typical data transformations include first differencing, that is, examining only the changes of demand between periods; calculating growth rates, that is, examining the normalized first difference; or taking the natural logarithm of the data. Suppose, for example, one observes the following four observations of a time series: 100, 120, 160, and 150. The corresponding three observations of the first difference series become 20, 40, and –10. Expressed as growth rates, this series of first differences becomes 20, 33, and –6%.

These transformations are reversible. While we make estimations on the transformed data, we can transform the resulting forecasts back to be useful for our purpose. The benefit of such transformations usually lies in the reduction of variability and in filtering out the unstable portions of the data. Statistical software generally reports several statistical tests for stationarity, such as the Dickey–Fuller test. Applying these tests to examine whether a first-differenced time series has achieved stationarity is helpful.

5.5 Forecastability and scale

Another aspect to understand about a time series is the forecastability of the series. As discussed in Chapter 1, some time series contain more noise than others, making predicting their future realizations more challenging. The less forecastable a time series is, the wider the prediction interval associated with the forecast will be. Understanding the forecastability of a series not only helps set expectations among decision-makers but is also essential when

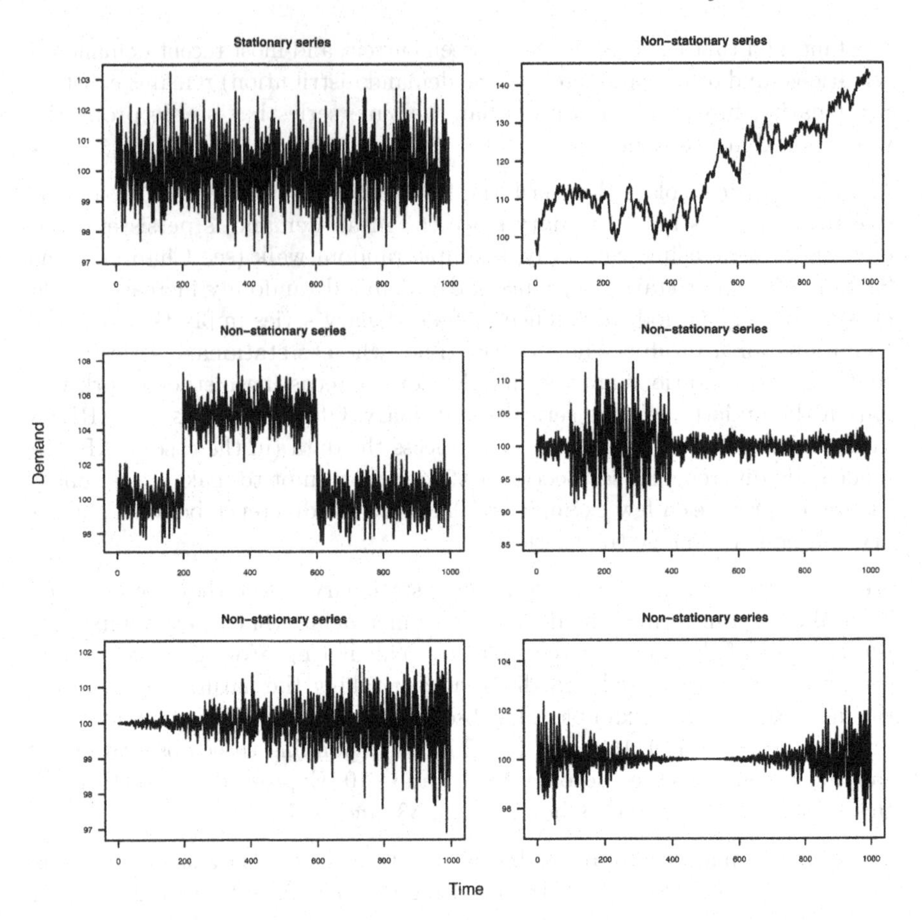

FIGURE 5.1
Examples of a stationary versus non-stationary time series

examining appropriate benchmarks for forecasting performance. Competitors' forecasts may be more accurate if they have a better forecasting process, or if their time series are more forecastable. The latter may simply be a result of them operating at a larger scale, with less variety, or their products being less influenced by current fashion and changing consumer trends.

One measure of the forecastability of a time series is the ratio of the standard deviation of the time series to the standard deviation of errors for forecasts calculated using a benchmark method as per Chapter 8 (Hill, Zhang, and Burch 2015). The logic behind this ratio is that the standard deviation of demand is, in some sense, a lower bound on performance since it generally corresponds to using a simple, long-run average as your forecasting method for the future. Any proper forecasting method should not lead to more uncertainty than the uncertainty inherent in demand. Thus, if this ratio is > 1, forecasting a time series may benefit from more complex methods than a long-run average. If this ratio is close to 1 (or even < 1), we may not be able to forecast the time series any better than using a long-run average.

This conceptualization resembles what some researchers call *Forecast Value Added* (Gilliland 2013). In this concept, one defines a base accuracy for a time series by calculating the forecast accuracy achieved (see Chapter 17) by the best simple method (see Chapter 8). Every step in the forecasting process, whether it is the output of a statistical forecasting model, the consensus forecast from a group, or the judgmental adjustment to a forecast by a higher-level executive, is then benchmarked in terms of their long-run error against this base accuracy. If a method requires effort from the organization but does not lead to better forecast accuracy compared to a method that requires less effort, we can safely eliminate it from future forecasting processes. Results from such comparisons are often sobering. Some estimates suggest that in almost 50% of time series, the existing toolset available for forecasting does not improve upon simple forecasting methods (Morlidge 2014a). In other words, demand averaging or simple demand chasing may sometimes be the best a forecaster can do to create predictions.

Some studies examine what drives the forecastability of a series (Schubert 2012). Key factors include the overall volume of sales (larger volume means more aggregation of demand, thus less observed noise), the coefficient of variation of the series (more variability relative to mean demand), and the intermittency of data (data with only a few customers who place large orders is more difficult to predict than data with many customers who place small orders). In a nutshell, we can explain the forecastability of a time series by product characteristics as well as firm characteristics within its industry. There are economies of scale in forecasting, with forecasting at higher volumes being easier than forecasting for low volumes.

The source of these economies of scale lies in the principle of statistical aggregation. Imagine trying to forecast who among all the people living in your street will buy a sweater this week. You would end up with a forecast for each person living in the street that is highly uncertain for each individual. However, the task becomes much easier if you just want to forecast how many people living in your street buy a sweater in total. You can make many errors at the individual level, but at the aggregate level, these errors cancel out. This effect will increase the more you aggregate, predicting at the neighborhood, city, county, state, region, or country level. Thus, the forecastability of a series is often a question of what level of aggregation a time series focuses at. Very disaggregate series can become intermittent and, therefore, very challenging to forecast (see Chapter 12 for details). Very aggregate series are easier to forecast, but if the level of aggregation is too high, these forecasts become less useful for planning purposes, as the information they contain is not detailed enough.

It is vital in this context to highlight the difference between relative and absolute comparisons in forecast accuracy. In absolute terms, a series at a higher level of aggregation will have more uncertainty than each individual series. Still, in relative terms, the uncertainty at the aggregate level will be less than the sum of the uncertainties at the lower level. Suppose you predict whether a person will buy a sweater or not. In that case, your absolute error is at most 1, whereas the maximum error of predicting how many people in your street buy a sweater or not depends on how many people live in your street; nevertheless, the sum of the errors you make at the individual level will be less than the error you make in the sum. For example, suppose five people live in your street, and we can order them by how far they live in the street (i.e., first house, second house, etc.). You predict that the first two residents will buy a sweater, whereas the last three do not. Your aggregate prediction is just that two residents buy a sweater. Suppose now that only the last two residents buy a sweater. Your forecast is 100% accurate at the aggregate level but only 20% accurate at the disaggregate level. Generally, the standard deviation of forecast errors at the aggregate level will be less than the sum of the standard deviations of forecast errors made at the disaggregate level.

Supply chain design can allow using more aggregate forecasts in planning. The benefits of aggregation here are not limited to better forecasting performance but also include reduced inventory costs. For example, the concept of postponement in supply chain design favors postponing the differentiation of products until later in the process. This practice enables forecasting and planning at higher levels of aggregation for longer within the supply chain. Paint companies were early adopters of this idea by producing generic colors that are mixed into the final product at the retail level. This postponement of differentiation allows forecasting (and stocking) at much higher levels of aggregation. Similarly, Hewlett-Packard demonstrated how to use distribution centers to localize their products. This allowed them to produce and ship generic printers to distribution centers.

A product design strategy that aims for better aggregation is component commonality or a so-called platform strategy. Here, components across stock-keeping units are kept in common, enabling production and procurement to operate with forecasts and plans at a higher level of aggregation. Volkswagen is famous for pushing the boundaries of this approach with its MQB platform, which allows component sharing and final assembly on the same line across such diverse cars as the Audi A3 and the Volkswagen Touran. Additive manufacturing may become a technology that allows planning at very aggregate levels (e.g., printing raw materials and flexible printing capacity), allowing companies to deliver various products without losing economies of scale in forecasting and inventory planning.

5.6 Using summary statistics to understand your data

Using summary statistics, such as the mean (average) and the standard deviation is an initial step to understand data, as discussed above. These measures may help to condense data, compare different datasets, and convey complex messages.

While summary statistics can provide a basic understanding of data, however, their usefulness for time series analysis is limited. First, they cannot highlight patterns in the data, identify unusual observations, describe changes over time, or help to understand relationships between variables. Second, they could be misleading. For example, a set of four datasets known as *Anscombe's Quartet*, as well as another collection of 13 datasets called the *Datasaurus Dozen*, each have the same summary statistics – means and standard deviations of both variables, as well as correlations – but are completely different when plotted (Matejka and Fitzmaurice 2017). Thus, similar statistics can stem from very different datasets. Therefore, we should never rely exclusively on summary statistics alone in our judgment and always visualize our data.

Visualising data can communicate information much quicker than summary statistics and provides a better tool for analyzing and understanding time series data. In Chapter 6 we will see how different time series graphics can be used to better understand data.

Key takeaways

1. Understanding your data is the first step to a good forecast.
2. The objective of most forecasts is to predict demand, yet the data available to prepare these forecasts often reflects sales; if stockouts occur, sales are less than demand.
3. Many forecasting methods require time series to be stationary, that is, to have constant parameters over time. We can often obtain stationarity by suitable transformations of the original time series, such as differencing the series.
4. A vital attribute of a time series is its forecastability. Your competitors may have more accurate forecasts because their forecasting process is better or their time series are more forecastable.
5. There are economies of scale in forecasting; predicting at a larger scale tends to be easier due to statistical aggregation effects.

6. Throwing away time series data just because it is "old" is not a good practice. The longer your time series, the better your model becomes.
7. Do not rely on summary statistics alone to understand your data; always plot your data.

6

Time series components

This chapter discusses basic time series components that a forecasting model must incorporate, like the level, the trend and seasonality. While the next chapter will discuss decomposing the time series into components more explicitly, this chapter provides a basic familiarity with these components.

Plotting a time series is an excellent first step to identifying these components. This chapter describes some typical plots useful for visualizing univariate time series data. Such graphs may also surprise you by revealing unexpected data structures.

6.1 Level

The *level* of a time series describes the center of the series. If we could imagine the time series without random noise, trend, and seasonality, the remainder would be the level. Using a time plot, we illustrate this thought process on the left-hand side of Figure 6.1. A time plot means that we plot the variable against time. The plot on the right-hand side of Figure 6.1 shows a time series with random variations around the level. This plot is more realistic because some randomness influences every observation.

6.2 Trend

A *trend* describes predictable increases or decreases in the level of a series. A time series that grows or declines in successive periods has a trend. We illustrate this time series component in Figure 6.2 with an increasing (left-hand) and decreasing (right-hand) trend. By definition, trends need to be somewhat stable to allow predictability. It is often challenging to differentiate a trend from abrupt shifts in the level, as discussed in Chapter 4. If one reasonably expects the level shift to reoccur similarly in the following period, one can speak of a trend instead of an abrupt level shift. A long-run persistence of

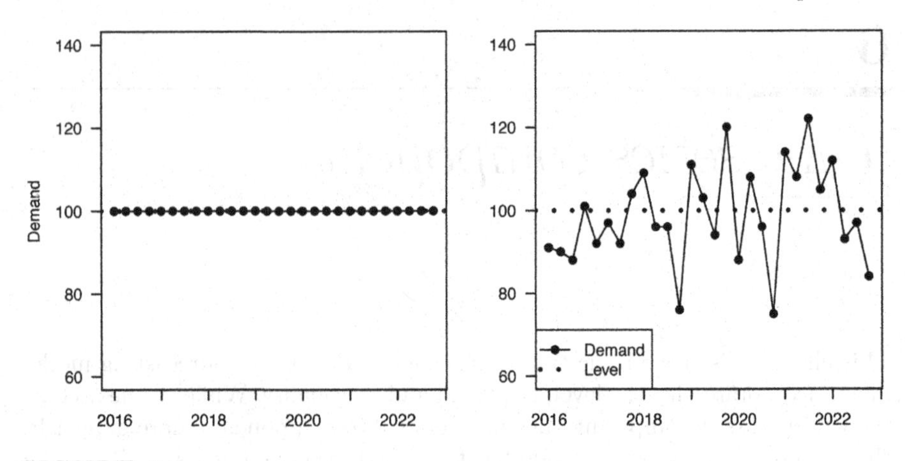

FIGURE 6.1
Time series with a level only (left), and a level and random noise (right)

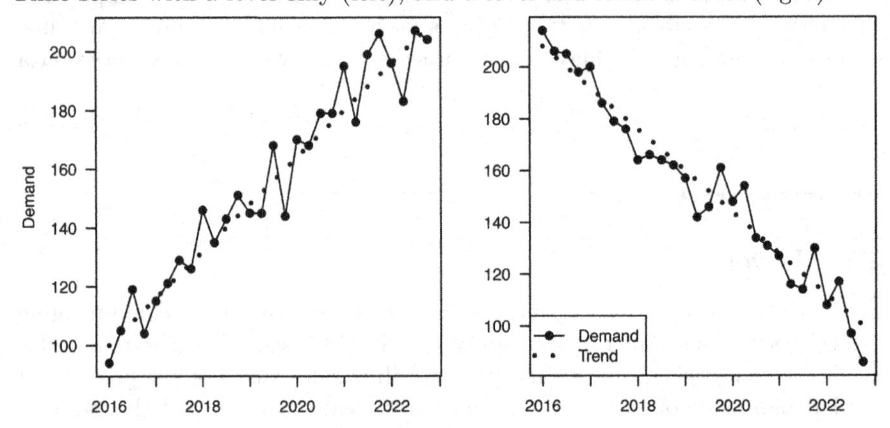

FIGURE 6.2
Time series with an increasing (left) and decreasing (right) trend

such increases/decreases in the data is necessary to establish that a real trend exists.

A trend in the time series can have many underlying causes. At the beginning of its lifecycle, a product will experience a positive trend as more and more customers receive information about the product and decide to buy it. On the upside of the business cycle, the gross domestic product expands, making consumers wealthier and more able to purchase. More fundamentally, the world population is growing by about 1% yearly. To some degree, a firm that sells products globally should observe this global population increase as a trend in their demand patterns.

Trends can also be either linear or non-linear. Linear trends are positive or negative additive increments to the series level. An additive trend implies a linear increase/decrease, that is, an increase/decrease in demand by X units in every period. Non-linear trends are often multiplicative, with increments proportional to the previous series value(s). A multiplicative trend implies an exponential increase/decrease, that is, an increase/decrease in demand by X *percent* in every period. Multiplicative trends tend to be easier to interpret since they correspond to statements like "our business grows by 10% every year"; however, if the trend does not change, such a statement implies unbounded exponential growth over time. We can expect such growth patterns in the early stages of a product lifecycle. Still, time series models using multiplicative trends must pay extra attention to not set such growth in stone but allow it to taper off over time.

6.3 Seasonality

Seasonality is a consistent pattern that repeats over a fixed cycle. For example, in a daily time series, the cycle may repeat itself every 7 days (a weekly seasonal pattern). Patterns over 7 days should look similar from week to week. Other examples of seasonality include predictable increases in demand for consumer products every December for the holiday season or higher demand for air-conditioning units or ice cream during the summer. A company's regular promotion event in May will also appear as seasonality.

The strength of seasonality often depends on the time granularity (see Section 15.1). Data on yearly granularity usually has little seasonality. While leap years create a regularity that reappears every 4 years by including an extra day in the year, this effect is small enough to be ignored. Temperature often influences data on sub-yearly granularity. Weekly or daily data can additionally be subject to day-of-the-week, payday, billing cycle, and end-of-month effects (Rickwalder 2006). Hourly data will often have visible time-of-the-day effects. Various factors, such as weather patterns, administrative measures, and social, cultural, and religious events, may cause seasonalities. Some calendar-related effects may change and fall in different months from year to year, such as Easter. All these effects are treated as seasonality in time series forecasting since they represent predictable recurring patterns over time. There can also be more complex seasonal patterns, which we address in Chapter 15.

Additive vs. multiplicative seasonality

Seasonality can appear in two forms: additive and multiplicative. In the former case, the magnitude of seasonality does not change relative to the series level.

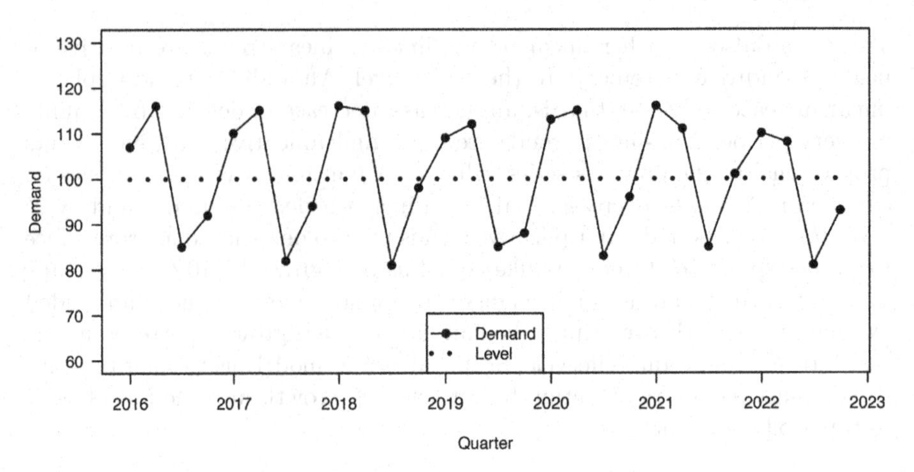

FIGURE 6.3
A time series with level and additive seasonality

In the latter, seasonal fluctuations increase or decrease proportionally with increases and decreases in the level. We illustrate additive and multiplicative seasonality in Figure 6.4. You will notice the difference in the peak and trough amplitudes. Specifically, the amplitude of the seasonal component of the multiplicative time series changes with the trend.

Additive seasonality usually applies to more mature products with relatively little growth. In contrast, multiplicative seasonality naturally incorporates growth in a series, particularly in contexts where the effect of seasonality depends on the scale of demand. Differentiating between additive and multiplicative forms of trend and seasonality is vital for the general Exponential Smoothing framework we discuss in Chapter 9.

Seasonal and seasonal subseries plots

Identifying the seasonal pattern by observing only a time series plot can be challenging. A *seasonal plot* (also known as a *seasonplot*) can be more helpful. In a seasonal plot, we plot the data against the "seasons" in which we observed them, with one line per full cycle. This plot allows us to spot the underlying seasonal pattern and possible changes in this pattern. We plot the same data as before in Figure 6.3 as a seasonal plot.

We also exhibit a *seasonal subseries plot*, which works a little differently. Here, we collect and plot observations based on their "season" (quarters 1 to 4 for quarterly data, weekdays for daily data) against the cycles in separate time sub-plots. We also show the mean for each season as a horizontal line. Such a plot is especially useful in identifying *changes* within seasons from cycle to cycle.

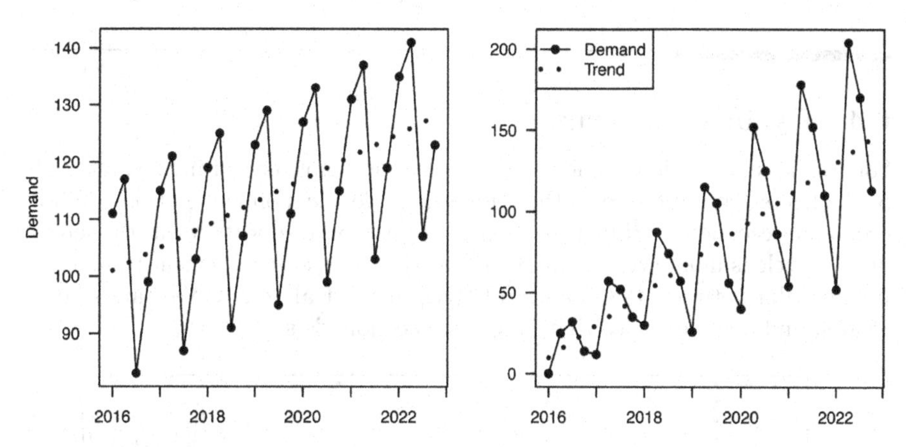

FIGURE 6.4
Time series with an increasing trend and additive (left) and multiplicative (right) seasonality

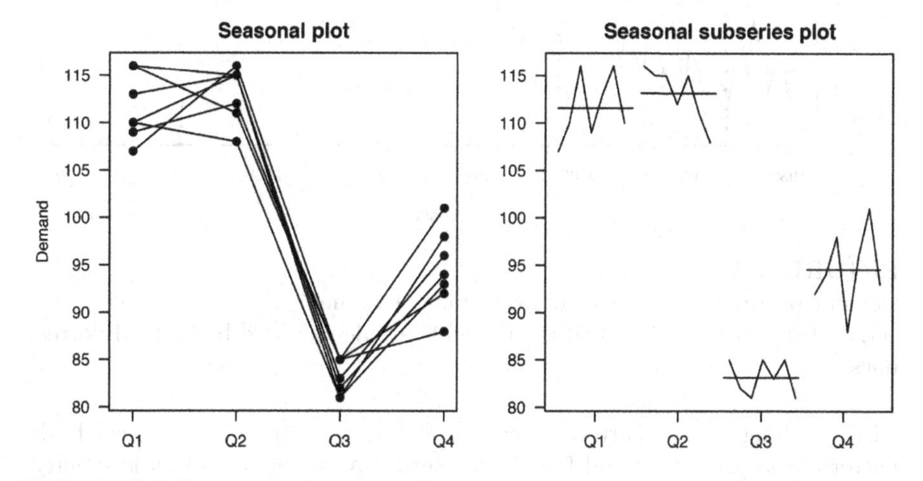

FIGURE 6.5
A seasonal (left) and a seasonal subseries (right) plot for the time series shown in Figure 6.3

We can generate seasonal plots and seasonal subseries plots with functions in the `forecast` (Hyndman et al. 2023) and `feasts` (O'Hara-Wild, Hyndman, and Wang 2022) packages for R.

6.4 Cyclical patterns

Some series may exhibit cyclical behavior that does not reoccur in fixed intervals. Such effects are often due to the economic cycle. Economic expansions follow economic recessions. While this boom and bust cycle repeats itself, the length of each cycle is unknown. Figure 6.6 shows the time series of cement production in Australia between 1980 and 2014 (Hyndman et al. 2023). We can observe growth and decline cycles that follow the economy's state.

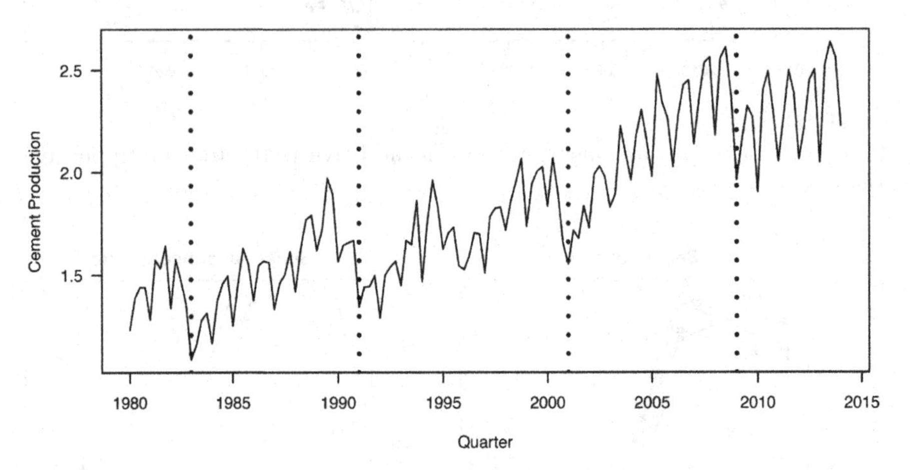

FIGURE 6.6
Cement production in Australia (millions of tonnes). Business cycles in the early 1980s, 1990s, 2000s, and around 2008 are demarcated by vertical dotted lines

Distinguishing cyclical patterns from seasonal ones can be confusing since both patterns comprise rises and falls in the series. A seasonal pattern is strictly regular, meaning that the distance between the peaks and troughs of the series is the same (e.g., every 4 quarters, 12 months, 7 days, 24 hours, etc.). Cyclical patterns are not as regular. They can drift over time, and the distance between rises and falls is not fixed. Cyclical patterns may occur over multiple years. Time series can combine both cyclical and seasonal patterns. For instance, the time series of cement production shown in Figure 6.6 has a cyclical effect due to the market conditions and a seasonal effect likely induced by the change in weather conditions, which drive seasonality in construction.

6.5 Other characteristics of time series

Another essential characteristic of a time series is the association between demand at the current period and previously observed demands. A scatterplot allows you to graph an observation against a prior observation. For example, you could plot demand in period t against demand in period $t - 1$. Such a scatterplot is also called a *lag plot*, because you are plotting the time series against lags of itself. A time series with lag 1 is a version of the original time series that is one period behind in time.

If a time series shows a correlation with a lagged version of itself, it is said to exhibit *autocorrelation* (from Greek *auto*, meaning "self"). The correlation between a series and itself lagged by 1 step is the *lag 1 autocorrelation*, and in a similar way, we can have lag 2, 3, etc. autocorrelations. An autocorrelation shows up as a correlation in lag plots like Figure 6.7.

Intuitively, autocorrelation implies that the current demand can help predict the immediate future. Thus, assessing autocorrelation and incorporating previous lags into models can lead to more accurate forecasts, especially for datasets where trends and seasonalities are difficult to determine precisely. We discuss ARIMA models that account for autocorrelation in Chapter 10.

Another characteristic that we see in a time series is randomness. Time series can consist of systematic patterns (i.e., trend, seasonality, autocorrelation) and random variations. For example, a time series of air pollution might have day-of-week effects based on traffic or commuting patterns and some randomness. Some time series rise and fall with no apparent trend, seasonality, or autocorrelation. Time series without any patterns at all are called *white noise*, meaning that the nature of the process generating the data is inherently random, unknown, and unpredictable (except possibly for an overall level).

6.6 Visualizing time series features

The previous section outlined how to understand the components of individual series. While plotting and inspecting time series is valid, this process can be cumbersome for large-scale (e.g., thousands or millions of time series) forecasting tasks. Instead of plotting the individual series, you can quantify their structure by computing, plotting and analyzing several numerical summary statistics called *time series features* (not to be confused with "features" in the sense of predictors or explanatory variables, see Chapter 11).

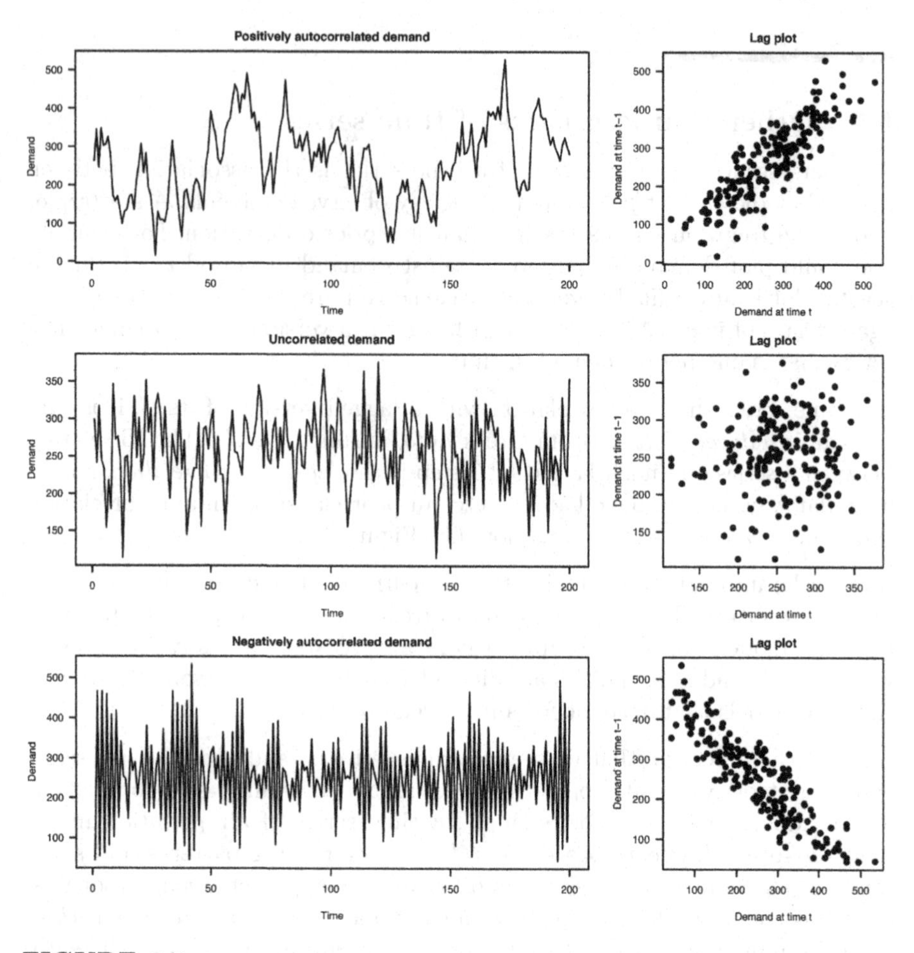

FIGURE 6.7
Lagged scatterplots for simulated time series with positive, zero or negative autocorrelation

We have already seen some time series features. We introduced basic features, such as the average (μ) and standard deviation (σ), in Chapter 6. Given these two features, you can compute the *coefficient of variation* (CV) as a new feature that measures the variability of a time series, where $\text{CV} = \sigma/\mu$. A $\text{CV} < 1$ indicates that the standard deviation is smaller than the average, while a $\text{CV} > 1$ indicates that the standard deviation is greater than the average. Higher values of CV correspond to high variability in the time series. We can use the CV as a measure of "forecastability," with a higher CV being in general less forecastable (see Section 5.5). Besides the CV, forecasters apply other features, like, e.g., the entropy, to quantify the forecastability of time series data instead (Kang, Hyndman, and Smith-Miles 2017).

Another time series feature we have previously discussed is the autocorrelation at various lags from Section 6.5. This feature captures the strength of the linear relationship associated with each lag plot as shown in Figure 6.7. Autocorrelations vary between -1 and $+1$. Autocorrelations closer to zero indicate little linear relationship between a time series and its previous lags. An autocorrelation coefficient close to -1 or $+1$ indicates a stronger linear association.

We can also compute the strength of the trend and the seasonality as features of a time series. Typical definitions calculate these strengths as a numerical value between 0 and 1 (Hyndman and Athanasopoulos 2021). A feature closer to 1 indicates a stronger trend and seasonality. These measures can be helpful when analyzing an extensive collection of time series.

We illustrate these features with the example of a dataset containing 1,224 monthly time series demand of three types of medical products in nine regions over the past couple of years. Plotting all these series using time series graphs is not practical. However, we can compute the CV for each series and plot a histogram of all these CVs to visualize the variability of our data. We illustrate this approach in Figure 6.8.

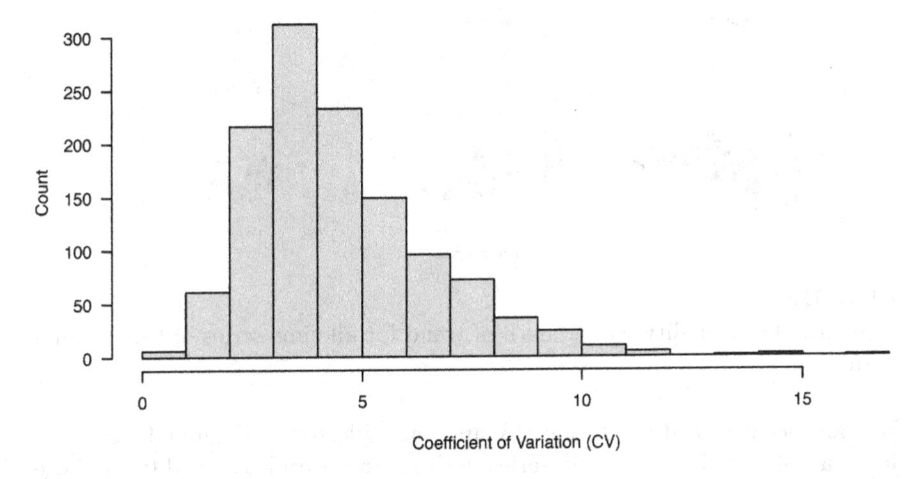

FIGURE 6.8
A histogram of coefficients of variation for 1,224 time series of medical product demand

The histogram reveals high variability in the medical product demand time series, indicating potential challenges in producing accurate forecasts. A few time series on the left show lower CVs, but most time series have high CVs.

As a next step, we can compute the strengths of trend and seasonality for each series and visualize them. Figure 6.9 is a scatterplot of seasonal strength versus trend strength. Each sub-plot contains time series for each region, and each point corresponds to one time series with different shapes corresponding to different types of medical products. Demand in most regions is not seasonal, except for Region F, where we can observe high seasonality for product type 1. Many regions exhibit a trend. Regions A, D, and E trends seem less strong compared to the other regions.

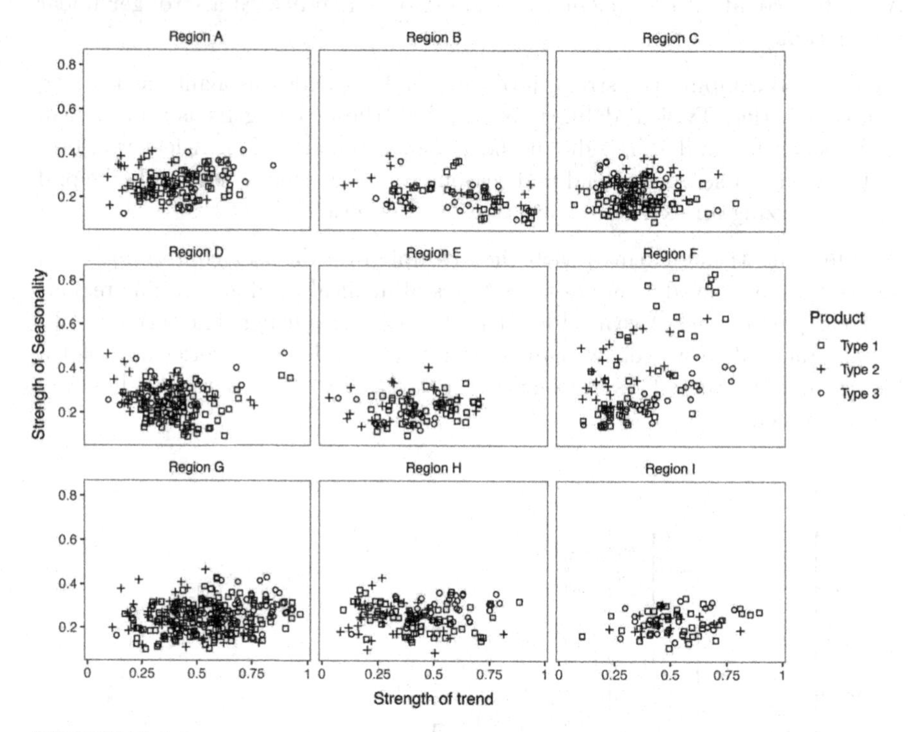

FIGURE 6.9
Strength of seasonality vs. strength of trend for all time series of the demand data

We can use these features to identify any series plotted in Figure 6.9. You can, for example, explore the time series with no trend and seasonality or those with the strongest trend or seasonality features. For instance, we identified a series with the highest seasonality in the medical product demand data, which belongs to product type 1 in region F. We visualize this series in Figure 6.10.

Key takeaways

1. Using a time plot is an excellent way to begin understanding your time series data.
2. Time series data can comprise systematic information, including level, trend, seasonality, and autocorrelation.

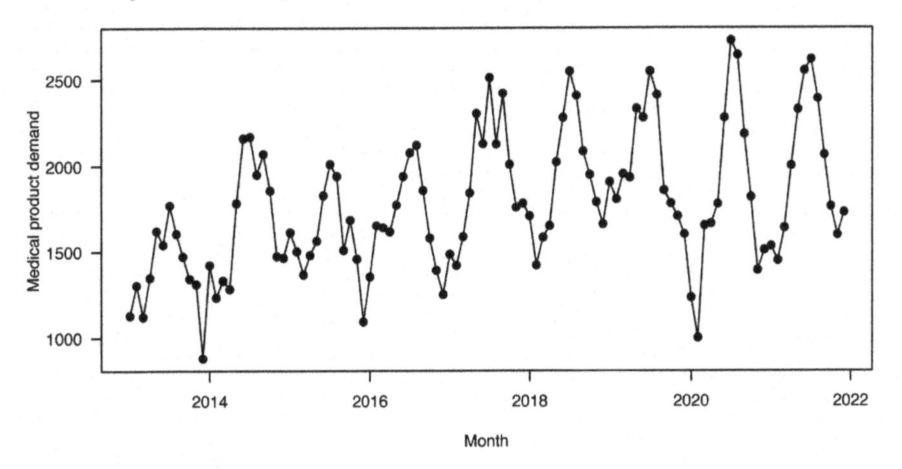

FIGURE 6.10
The time series with the strongest seasonality among the medical product demand series

3. The trend and seasonality patterns can be additive or multiplicative. Multiplicative components increase growth/decline patterns over time, whereas additive components imply a more linear growth/decline of firms.
4. Seasonal and seasonal subseries plots are helpful graphs to reveal seasonality.
5. Lag plots are an essential tool to understand the association between current and previous observations in time series, i.e., autocorrelation.
6. Unpredictable randomness is an inherent part of any time series.
7. When analyzing and forecasting an extensive collection of time series, extracting and plotting numerical time series features helps uncover useful information.

7

Time series decomposition

A very simple but often surprisingly effective way to forecast is to decompose a series into the components we encountered in the previous chapter, forecast each component separately, and combine the forecasts back together.

7.1 The purpose of decomposition

Decomposition is a management technique for complexity reduction. Separating a problem into its components and solving them independently before reassembling them into a broader decision enables better decision-making (e.g., Raiffa 1968).

The decomposition of univariate time series data is a standard tool many organizations use. It creates a cleaner picture of our time series, allowing us to better spot the tree within a forest. Here is why this can be helpful:

1. Decomposition provides a clean way to identify the trend and seasonality components. We may suspect that the trend requires dampening or that seasonality changes over time. Decomposition helps to check your assumptions.
2. Decomposition allows us to identify your series' causal drivers. For example, holidays or a football match can impact demand. Seasonality can obfuscate the effect of such events in our data. Removing seasonality through decomposition will thus allow us to identify the impact of such special events more easily.
3. Decomposition may alert us that two different types of seasonality influence your data (see Chapter 15). For example, when looking at daily data, the effects of the weekday may become more salient once we remove the monthly impact.
4. Decomposition enables us to spot outliers. Time series may contain outliers (i.e., anomalies) that are difficult to spot. Since

decomposition removes systematic variation, outliers become easier to detect in the remainder.

5. Decomposition allows us to create a seasonally adjusted series. For instance, governments want to understand and communicate the unemployment rate over time. Unemployment can be very seasonal, for example, due to a winter construction work slowdown. If we do not seasonally adjust a reported time series, readers might misinterpret an increase in the unemployment rate as substantive rather than driven by the season.
6. Decomposition can be used in forecasting. The technique provides a structured way of thinking about a time series forecasting problem. The key idea is to *decompose* a time series into separate components that can be examined, estimated, and forecasted separately. We can then recombine the component forecasts into a regular forecast. Many forecasting methods require data that has already been decomposed or incorporate a decomposition approach directly into the method.

7.2 Decomposition methods

We can use a decomposition method to decompose the time series data into trend, seasonality, and remainder. To reconstruct the data, these components can be added or multiplied. In the former case, we get data = trend+seasonal+ remainder, and in the latter, data = trend × seasonal × remainder. Figure 7.1 illustrates an original time series at the top that is decomposed into trend, seasonal, and remainder components using an additive method.

The remainder is whatever remains in the time series after removing trend and seasonality. If trend and seasonality capture all systematic variation in the data, then the remainder will only contain random noise. Otherwise, it will also include some systematic information, so any pattern visible in the remainder series can be very informative.

Historically, people tried to model the business cycle as an additional component. But since estimating the business cycle and predicting when the cycle turns is inherently very challenging, many methods nowadays do not explicitly consider the business cycle as a separate component. In addition, the trend component often captures the business cycle sufficiently well.

When interpreting publicly available time series, such as data from the Bureau of Labor Statistics, one needs to carefully understand whether or not seasonality has already been removed from the data. Most government data is reported as *seasonally adjusted*, implying that the seasonal component has been removed

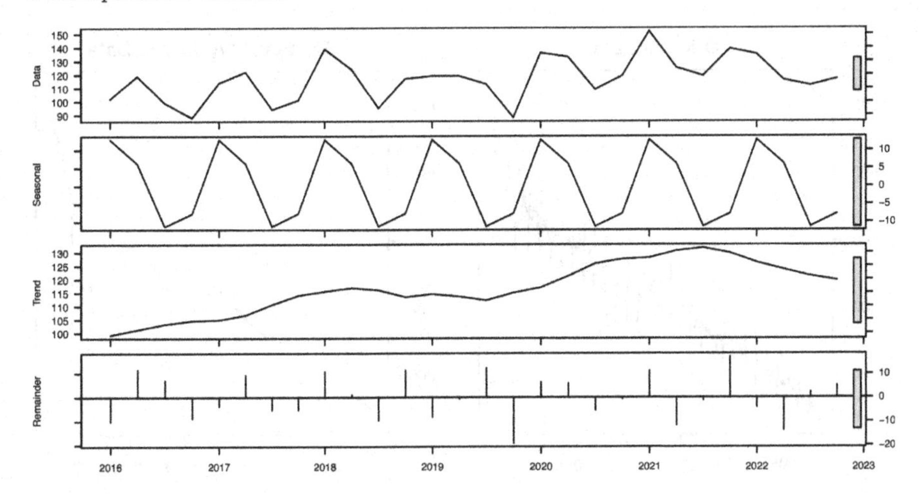

FIGURE 7.1
An original quarterly time series (top) along with the estimated seasonality, trend, and remainder

from the series. This decomposition is usually applied to the time series to avoid readers over-interpreting month-to-month seasonal changes.

Most commercial software offers methods to remove seasonality and trends from a time series. Taking seasonality and trends out of a time series is also easy in spreadsheet modeling software such as Microsoft Excel. Take monthly data as an example. As a first step, calculate the average demand over all data points and then calculate the average demand for each month (i.e., average demand in January, February, etc.). Dividing the average monthly demand by the overall average demand creates a *seasonal index*. Then dividing all demand observations in the time series by the corresponding seasonal index creates a deseasonalized series. We illustrate the results of such a seasonal adjustment in Figure 7.2.

Similarly, data can be *detrended* by first calculating the average difference between successive observations of the series and then subtracting $(n-1)$ times this average from the nth observation in the time series. For example, take the following time series: 100, 120, 140, and 160. The first differences of the series are 20, 20, and 20, with an average of 20. The detrended series is thus $100 - 0 \times 20 = 100$, $120 - 1 \times 20 = 100$, $140 - 2 \times 20 - 100$, and $160 - 3 \times 20 = 100$.

While these methods of deseasonalizing and detrending data are simple to use and understand, they suffer several drawbacks. For instance, they do not allow seasonality indices and trends to change, making their application challenging for more extended time series. Figure 7.2 illustrates an example of this problem. While initially, our method of deseasonalizing the data removed seasonality

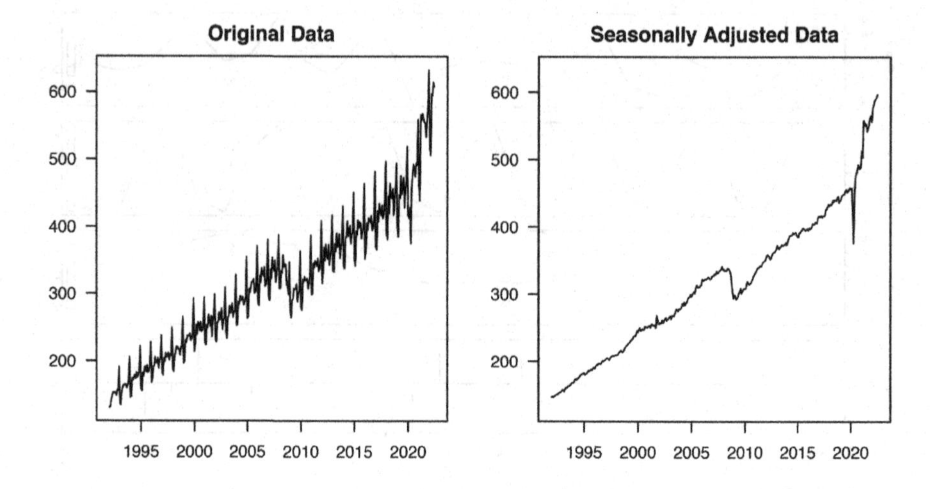

FIGURE 7.2
Monthly US retail sales 1992–2022 (Source: US Census Bureau, https://www.census.gov/retail/index.html)

from the series, the later part of the series exhibits a seasonal pattern again since the seasonal indices have changed. In shorter time series, outlier data can also strongly influence decomposition methods. More sophisticated (and complex) methods have been developed for these reasons. The Bureau of Labor Statistics has developed an algorithm called X-13ARIMA-SEATS (US Census Bureau, n.d.). Software implementing this algorithm is available to download for free from the Bureau's website.

Most time series decomposition methods are designed for monthly and quarterly data, with few tools available for daily and sub-daily series. For instance, hospitals record hourly patient arrivals in their services, typically showing a time-of-day pattern, a day-of-week pattern, and a time-of-year pattern (see Chapter 15). These seasonalities may also interact with special events such as holidays and sports events. One time series decomposition method for such high-frequency time series (e.g., hourly and daily) is the STL (seasonal-trend decomposition based on loess) procedure (Cleveland et al. 1990). An extension of the STL procedure, Multiple Seasonal-Trend decomposition using Loess (MSTL) can be used to decompose time series with multiple seasonal patterns (Bandara et al. 2021).

7.3 Stability of components

The observation that seasonal components can change leads to a critical discussion. The real challenge of time series analysis lies in understanding the *stability* of the components of a series. A perfectly stable time series is a series where the components do not change as time progresses – the level only increases through the trend, and the trend remains constant. The seasonality remains the same from year to year. If true, the best time series forecasting method works with long-run averages. Yet time series often are inherently unstable, and components change over time. The level of a series can abruptly shift as new competitors enter the market. The trend of a series can evolve as the product moves through its lifecycle. Even the seasonality of a series can change if underlying consumption or promotion patterns shift throughout the year.

To illustrate what change means for a time series and the implications that change has for time series forecasting, consider the two illustrative and artificially constructed time series in Figure 7.3. Both are time series without trends and seasonality (or deseasonalized and detrended already). Series 1 is a perfectly stable series, such that month-to-month variation resembles only random noise. This example comes from a stationary demand distribution and is typical for mature products. Series 2 is a time series that is highly unstable, such that month-to-month variation resembles only change in the underlying level. This example of data stems from a so-called *random walk* and is typical for prices in an efficient market. We used the same random draws to construct both series; in series 1, randomness represents just random noise around a stable level (around 500 units). The best forecast for this series would be a long-run average (i.e., 500). In series 2, randomness represents random changes in the unstable level, which can push the level of the time series up or down by the nature of randomness. The best forecast in this series would be the most recent demand observation. The same randomness can create two very different types of time series requiring very different approaches to forecasting.

In other words, uncertain but stable time series can use all available data to estimate the time series component and create a forecast. In unstable time series, we place a lot of weight on very recent data, and very distal data has very little weight in generating forecasts. Differentiating between stable and unstable components, and thus using or discounting past data, is the fundamental principle underlying Exponential Smoothing, which is a technique we will introduce in Chapter 9, while the "extreme" cases of purely stable or unstable series motivate the consideration of simple forecasting methods (see Chapter 8).

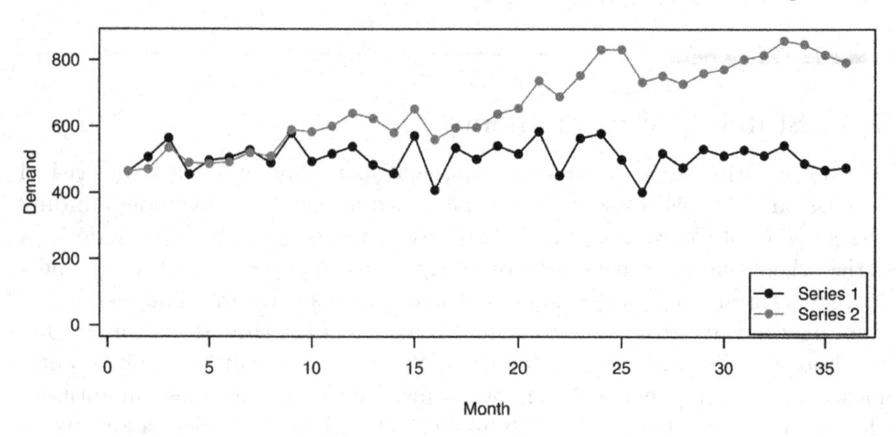

FIGURE 7.3
A stable and an unstable time series

Key takeaways

1. In time series decomposition, we separate a series into its seasonal, trend, level, and remainder components. We analyze these components separately and finally combine the pieces to create a forecast.
2. Components can be decomposed and recombined additively or multiplicatively.
3. Multiplicative components increase growth/decline patterns over time, whereas additive components imply a more linear growth/decline.
4. The challenge of time series modeling lies in understanding how much the time series components change over time.

Forecasting models

8

Low-hanging fruit: simple forecasts

Simple forecasting methods should be considered first. These methods are intuitive, and they can correspond to simple rules of thumb. For example, if you stock up to a level based on the average of the last 12 months of demand, you forecast demand using a rolling mean. We will discuss several such methods in this chapter.

There are three reasons for starting with simple methods:

1. They may give you a quick win without creating a complicated statistical forecast.
2. They give you a baseline to compare to more complicated forecasting methods.
3. They are often very effective. Morlidge (2014b) found that across multiple businesses, half of the forecasts did not improve on the naive no-change forecast.

8.1 The historical mean

One baseline benchmark every forecaster should look at is the simple *historical mean*: take all historical observations and average them. The result is your forecast for the entire future horizon. Simple, easy to understand and implement – and often surprisingly effective (Andy T 2016).

There is a statistical reason for its good performance. This simple forecast is optimal for a *white noise* statistical process, i.e., demand that randomly deviates around a long-term level.

8.2 The naive forecast

The *naive forecast*, also called the *no-change* or *random walk* forecast, means forecasting the last observation across the future. This is very simple and easy to understand and execute. Such a forecast will vary more from day to day than the historical mean: each time we see a new observation, the naive forecast can change radically. In contrast, the historical mean will be much less influenced by a single new observation. You can call this property of the naive forecast "high adaptivity" and consider it a feature or call it "high variance" and consider it a bug – which perspective is more helpful depends on your forecasting environment.

The naive forecast does have some statistical support. It is optimal for an integrated white noise process of order 1, or an I(1) process for short, which is a *random walk* – hence the name of this forecast. See Chapter 10 for details. Some measures to track forecast accuracy, such as the Mean Absolute Scaled Error (MASE), explicitly use this naive forecast as a baseline (see Chapter 17).

8.3 The seasonal naive forecast

If you predict a demand time series with strong seasonality, the naive forecast will always lag the seasonal signal. The historical mean, on the other hand, will smooth out any seasonality, pretending that it represents noise, for a forecast that is too low during the high season and too high during the low season.

A simple method that works for strongly seasonal series is the *seasonal naive forecast*, which is also referred to as a *seasonal no-change* or *seasonal random walk* forecast. This forecast is the most recent observation from the same point in the seasonal cycle. For example, our forecast for next January would be our observation from last January, and our forecast for next Wednesday would be our observation from last Wednesday.

Using this simple forecasting method also has a statistical justification: the seasonal naive forecast is optimal for a seasonally integrated process of order 1, i.e., one where an observation only depends on the observation one seasonal cycle before, plus random noise.

You may suspect that multiple seasonalities drive your data. For example, daily retail demand may exhibit weekly seasonality (higher demand on Friday and Saturday than during the rest of the week) and yearly seasonality (higher demand in summer than in winter). We will address more complex models that allow considering multiple seasonalities in Chapter 15 – here, we focus on simple methods. How should we create a naive forecast for a Wednesday in winter in our example? As a rule of thumb, you can pick the dominating seasonality for your simple forecast. If the fluctuations between days of the week are more substantial than the fluctuations over the year, take the demand from last Wednesday as a forecast for this Wednesday. And if the yearly fluctuations dominate, take the average demand from the same week of the previous year as a forecast. Of course, you can also use the average of these two possibilities (see Section 8.6 on the surprising effectiveness of averaging forecasts) or pick a historical Wednesday from roughly the same time last year as a forecast.

8.4 Other simple methods

Don't feel constrained by the three methods outlined above. There are other possibilities, like taking rolling averages of the last few observations. If you have multiple reasonable simple forecasts, it is often helpful to average them (see Section 8.6). The key is to do the simplest thing that could work, to profit from the three advantages given at the beginning of this chapter.

8.5 Non-expectation forecasts

So far, we have assumed that you want a point forecast that is right "on average." But recall that it is often helpful to understand more about the probability distribution of demand (see Section 3). We want to understand best-case and worst-case scenarios. We need to know how far the distribution spans around the center. Such an understanding is necessary to plan safety stocks and service levels effectively. How can we calculate prediction intervals or quantile forecasts in a simple way?

The simple methods above only provide quantile forecasts with tweaks. There are two simple ways forward. One approach is to calculate the demand time series's standard deviation and add one or two standard deviations to the central forecast to obtain an 85% or 97% quantile forecast. The other option would be to derive a simple quantile forecast from the time series using historical observations. For example, if we have 100 historical observations and want a 95% quantile forecast, we use the fifth-largest historical observation as our quantile forecast. In principle, such methods can factor in seasonality by only using data from the proper seasonal periods. In practice, getting reliable quantiles from seasonal data often requires too much data history.

8.6 Ensemble forecasting

One tool that often works quite well is taking the average of different possible forecasts. This is known as *ensemble forecasting* (from the French *ensemble*, meaning "together") or *forecast combination*, and there is much evidence that combining forecasts from different methods provides superior forecasts compared to the forecasts from the individual methods by themselves (Armstrong 2001). If we apply different methods to a time series, they will all pick up on different aspects or features of the data and extrapolate that into the future. Ensembling the forecasts together leverages all of them and adds stability, to boot. This approach works with both expectation and quantile forecasts. We can also add more complex or judgmental forecasting methods into the mix (see Section 16.4).

Straightforward ensembling takes the simple average of the constituent forecasts. All forecasts receive the same weight in the ensemble. A logical extension would be to create an ensemble with optimized weights. For instance, we could derive weights from how well the constituent forecasts worked in the past. Somewhat suprisingly, this idea does not work as well as one would expect. The fact that an unweighted average forecast often beats a weighted average using optimized weights has been referred to as the *forecast combination puzzle*. One explanation for it, proposed by Claeskens et al. (2016), is that the optimization of weights introduces uncertainty or noise into the entire forecasting pipeline – and this noise directly carries through to more noisy forecasts.

One can remove forecasts that do not add value (assuming we can identify them reliably) and create an ensemble from the remaining forecasts. This has been referred to as *forecast pooling* (Kourentzes, Barrow, and Petropoulos 2019).

We conclude this section with an important warning: do not attempt to build complex models *instead* of ensembles. Ensembling works because the constituent forecasts all pick up on and extrapolate different features of the underlying time series into the future. If that is so, why don't we build one complex model that accounts for all possible features by itself instead? This approach is not recommended, since it requires an enormous amount of data. See Section 11.7 for an explanation.

8.7 When are our forecasts *too* simple?

Simple methods are a good beginning. There are two cases where simple methods will surely be insufficient:

1. If strong drivers underlie your time series, we need to measure and account for them in our model (see Chapter 11). Such drivers can be anything, from promotions to the state of the economy. However, even in the presence of such drivers, you may be able to use a simple benchmark method, like using the sales from the last promotion as a forecast for your next promotion.
2. If added accuracy adds a lot of business value, investing money and resources into improved data and better models may be worthwhile. Just be aware that in light of irreducible noise, we may not be able to achieve the accuracy we would like to get with commensurate effort (see Chapter 21).

Thus, KISS: Keep It Sophisticatedly Simple.

Key takeaways

1. Simple forecasting methods are often surprisingly accurate – and simple to implement and explain.
2. You should always implement simple methods, at least as a sanity check or a baseline.
3. The most common simple methods are the historical mean, the naive forecast, and the seasonal naive forecast.
4. A simple recipe that often improves accuracy is to average different forecasts.
5. If you have *important* drivers, include them. KISS: Keep It Sophisticatedly Simple.

9

Exponential smoothing

One of the most accurate and robust methods for time series forecasting is Exponential Smoothing. Its performance advantage was established in the M3 forecasting competition, which compared more than 20 forecasting methods for their performance across 3,003 time series from different industries and contexts (Makridakis and Hibon 2000). You might think that in the 20 years since the M3 competition, more modern methods, like the ones based on Artificial Intelligence and Machine Learning (see Chapter 14), would have caught up to this venerable method. Surprisingly, this is not the case. Exponential Smoothing was still competitive in the more recent M5 competition, outperforming 92.5% of all submissions (Makridakis, Spiliotis, and Assimakopoulos 2022; Kolassa 2022a). In a world of rapid innovation and research, a method that has been around at least since the 1950s still shines.

9.1 Change and noise

There is extensive theoretical work on Exponential Smoothing. The method is optimal (i.e., it will minimize expected squared forecast errors) for random walks with noise and various other time series models (Chatfield et al. 2001; Gardner 2006). Further, what was once a rather ad-hoc methodology is now formalized in a state space framework (Hyndman et al. 2008). Exponential Smoothing is not only an accurate, versatile, and robust method. It is also intuitive and easy to understand and interpret. Another important aspect of the method is that the data storage and computational requirements for Exponential Smoothing are minimal. We can easily apply this method to many time series in real time.

To illustrate how Exponential Smoothing works, we begin with a simple time series without trend and seasonality. In such a series, variance in demand from period to period is driven either by random changes to the level (i.e., long-term shocks to the time series) or by random noise (i.e., short-term shocks to the time series). The essential forecasting task then becomes estimating the level of the series in each period and using that level as a forecast for the next period.

The key to effectively estimating the level in each period is to differentiate level changes from random noise (i.e., long-term shocks from short-term shocks).

As discussed in Section 7.3, if we believe that there are no random-level changes in the time series (i.e., the series is stable), then our best estimate of the level involves using all of our available data by calculating a long-run average over the whole time series. If we believe there is no random noise in the series (i.e., we can observe the level), then our estimate of the level is simply the most recent observation, discounting everything that happens further in the past. Not surprisingly, the *right* thing to do in a time series containing both random-level changes and random noise is something in between these two extremes: calculate a weighted average over all available data, with weights decreasing the further we go back in time. This approach to forecasting, in essence, is Exponential Smoothing.

Let the index t describe the period of a time series. We update the level estimate (= *Level*) in each period t according to Exponential Smoothing as follows:

$$Level_t = Level_{t-1} + \alpha \times Forecast\ Error_t. \tag{9.1}$$

In this equation, the coefficient α (alpha) is a smoothing parameter and lies somewhere between 0 and 1 (we will pick up the topic of what value to choose for α later in this chapter). The Forecast Error in period t is simply the difference between the actual demand (= Demand) in period t and the forecast made for the period t:

$$Forecast\ Error_t = Demand_t - Forecast_t. \tag{9.2}$$

In other words, Exponential Smoothing follows the simple logic of feedback and response. The forecaster estimates the current time series level, which they use as the forecast. This forecast is then compared to the actual demand in the series in the next period. This assessment allows the forecaster to revise the level estimate according to the discrepancy between the forecast and actual demand. The forecast for the next period is then simply the current level estimate since we assumed a time series without trend and seasonality, that is:

$$Forecast_{t+1} = Level_t. \tag{9.3}$$

Substituting Equations (9.2) and (9.3) into Equation (9.1), we obtain

$$\begin{aligned} Level_t &= Forecast_t + \alpha \times (Demand_t - Forecast_t) & (9.4)\\ &= (1-\alpha) \times Forecast_t + \alpha \times Demand_t. & (9.5) \end{aligned}$$

We can, therefore, also interpret Exponential Smoothing as a weighted average between our previous forecasts and the currently observed demand.

Curious readers have probably noticed that the method suffers from a chicken-or-egg problem: Creating an Exponential Smoothing forecast requires a previous forecast, which naturally creates the question of how to initialize the level component. Different initializations are possible, ranging from the first observed demand data point, an average of the first few demand points, to the overall average demand.

We can convert Equation (9.5) into a weighted average over all past demand observations, where the weight of a demand observation that is i periods away from the present is as follows:

$$Weight_i = \alpha \times (1 - \alpha)^i \tag{9.6}$$

Equation (9.6) implies an exponential decay in the weight attached to a particular demand observation the further this observation lies in the past. The technique is called Exponential Smoothing because of exponential decay. Figure 9.1 shows the weights assigned to past demand observations for typical values of α. We see that higher values of α yield weight curves that drop faster as we go into the past. The more recent history receives more weight than the more distant past if α is higher. Thus, forecasts will be more adaptive to changes for higher values of α. Note that the weight assigned under Exponential Smoothing to a period in the distant past can become very small (particularly with a high α) but remains positive nonetheless. We further discuss how to set the value of α in Section 9.2 below.

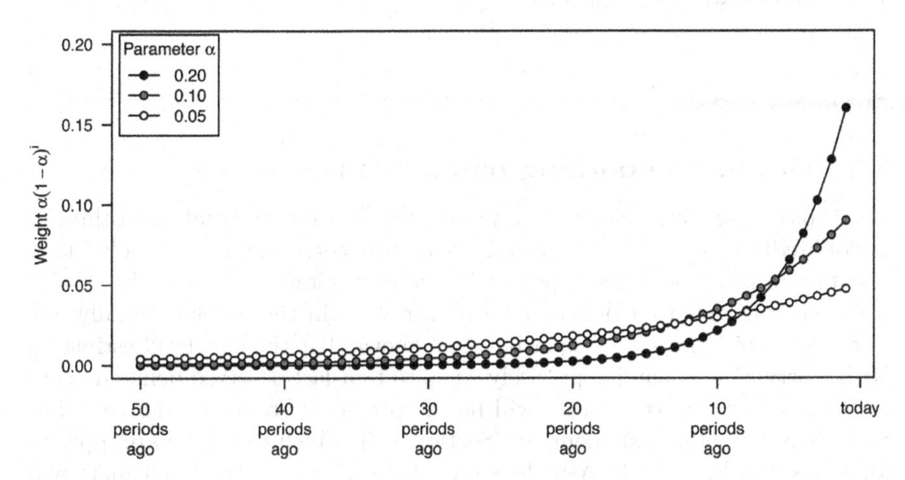

FIGURE 9.1
Weights for past demand under Exponential Smoothing

Forecasts created through Exponential Smoothing can thus be considered a weighted average of all past demand data. The weight associated with each demand decays exponentially the more distant a demand observation is from the present. It is essential, though, that one does not have to calculate a weighted average over all past demand to apply Exponential Smoothing. As long as a forecaster consistently follows Equation (9.1), all that is needed is a memory of the most recent forecast and an observation of actual demand to update the level estimate. Therefore, the data storage and retrieval requirements for Exponential Smoothing are extremely low. Consistently applying the method generates forecasts as if one were calculating a weighted average over all past demand with exponentially decaying weights in each period, as shown in Figure 9.1.

Forecasters sometimes use *moving averages* as an alternative to Exponential Smoothing. In a moving average of size n, the most recent n demand observations are averaged with equal weights to create a forecast for the future. This method is essentially similar to calculating a weighted average over all past demand observations, with the most recent n observations receiving equal weights and all others receiving zero weight.

However, why should there be such a step change in weighing past demands? Under Exponential Smoothing, all past demand observations receive some weight. This weight decays the more you move into the past. But moving averages assign equal weights for a while and no weight to the more extended demand history. An observation, e.g., six periods ago, might receive the same weight as the most recent observation, but an observation seven periods ago receives no weight. This radical change in weights from one period to the next makes moving averages appear arbitrary.

9.2 Optimal smoothing parameters

The choice of the *right* smoothing parameter α in Exponential Smoothing is undoubtedly essential. Conceptually, a high α corresponds to a belief that variation in the time series is primarily due to random changes in the level; a low α corresponds to a belief that this variation in the series is mainly due to random noise. If $\alpha = 1$, Equation (9.5) shows that the new level estimated after observing a demand is precisely equal to that last observed demand. That is, for $\alpha = 1$, the next forecast will be the previously observed demand, i.e., we have a naive forecast (compare Section 8.2), which would be appropriate for series 2 in Figure 7.3. As α becomes smaller, the new level estimate will become more and more similar to a long-run average. A long-run average would be correct for a stable time series such as series 1 in Figure 7.3. In the extreme, if $\alpha = 0$, per Equation (9.1) we will *never* update our estimate of the level, so the forecast will be completely non-adaptive and only depend on how we originally initialized this Level variable.

Choosing the right α thus corresponds to selecting the type of time series that our focal time series resembles more; series that look more like series 1 in Figure 7.3 should receive a higher α; series that look more like series 2 should receive a lower α. In practice, forecasters do not need to make this choice. They can rely on optimization procedures that will fit Exponential Smoothing models to past data, minimizing the discrepancy between hypothetical past forecasts and actual past demand observations by changing the α. The output of this optimization is a good smoothing parameter to use for forecasting. We can interpret the smoothing parameter as a measure of how much the demand environment in the past has changed over time. High values of α mean the market was very volatile and constantly changing, whereas low values of α imply a relatively stable market with little persistent change. In demand forecasting, smaller values of α, up to about $\alpha = 0.2$, typically work well since demand usually does have an underlying level that is relatively stable. In contrast, stock prices typically exhibit more random-walk-like behavior, corresponding to higher α.

9.3 Extensions

The simple incarnation of Exponential Smoothing described in Section 9.1 is often called *Single Exponential Smoothing (SES)*, or *Simple Exponential Smoothing* because it is the simplest version of Exponential Smoothing one can use, and because it has a single component, the level, and a single smoothing parameter α. One can extend the logic of Exponential Smoothing to forecast more complex time series. Take, for example, a time series with an (additive) trend and no seasonality. In this case, one proceeds with updating the level estimate per Equation (9.1), but then additionally estimates the trend in period t as follows:

$$Trend_t = (1 - \beta) \times Trend_{t-1} + \beta \times (Level_t - Level_{t-1}) \tag{9.7}$$

This method is also referred to as *Holt's method.* Notice the similarity between Equations (9.5) and (9.7). In Equation (9.7), beta (β) is another smoothing parameter that captures the degree to which a forecaster believes that the trend of a time series is changing. A high β would indicate a trend that can rapidly change over time, and a low β would correspond to a more or less stable trend. We again need to initialize the trend component to start the forecasting method, for example, by taking the difference between the first two demands or calculating the average trend over the entire demand history. Given that we assumed an additive trend, the resulting one-period-ahead forecast is then provided by

$$Forecast_{t+1} = Level_t + Trend_t \tag{9.8}$$

Usually, forecasters need not only to predict one period ahead into the future, but longer forecasting horizons are necessary for successful planning. In production planning, for example, the required forecast horizon is given by the maximum lead time among all suppliers for a product – often as far out as 6 to 8 months. Exponential Smoothing easily extends to forecasts further into the future as well; for example, in this case of a model with additive trend and no seasonality, the h-step-ahead forecast into the future is calculated as follows:

$$Forecast_{t+h} = Level_t + h \times Trend_t \tag{9.9}$$

This discussion requires emphasizing a critical insight and common mistake for those not versed in applying forecasting methods: Estimates are only updated if new information is available. We cannot meaningfully update level or trend estimates after making the one-step-ahead forecast and before making the two-step-ahead forecast. Thus, we use the same level and trend estimates to project the one-step-ahead and the two-step-ahead forecasts at period t. In the extreme, in smoothing models without a trend or seasonality, this means that all forecasts for future periods are the same; that is, if we expect the time series to be a *level only* time series without a trend or seasonality, we estimate the level once. That estimate becomes our best guess for demand in all future periods.

Of course, we understand that this forecast will get less accurate the more we project into the future, owing to the potential instability of the time series (a topic we have explored already in Chapter 4). Still, this expectation does not change the fact that we cannot derive a better estimate of what happens further in the future at the current time. After observing the following period, new data becomes available, and we can again update our estimates of the level and, thus, our forecasts for the future. Therefore, the two-step-ahead forecast made in period t (for period $t + 2$) usually differs from the one-step-ahead forecast made in period $t + 1$ (for the same period $t + 2$).

A trended Exponential Smoothing forecast extrapolates trends indefinitely. This aspect of the model could be unrealistic for long-range forecasts. For instance, if we detect a negative (downward) trend, extrapolating this trend out will eventually yield negative demand forecasts. Or assume that we are forecasting market penetration and find an upward trend – in this case, if we forecast out far enough, we will get forecasts of market penetration above 100%. Thus, we always need to truncate trended forecasts appropriately. In addition, few processes grow without bounds, and it is often better to *dampen* the trend component as we project it into the future. Such trend dampening will not make a big difference for short-range forecasting but will strongly influence, and typically improve, long-range forecasting (Gardner and Mckenzie 1985).

Similar extensions allow Exponential Smoothing to apply to time series with seasonality. Seasonality parameters are estimated separately from the trend and level components. We use an additional smoothing parameter γ (gamma) to reflect the degree of confidence a forecaster has that the seasonality in the time series remains stable over time. With additive trend and seasonality, we refer to such an Exponential Smoothing approach as *Holt–Winters Exponential Smoothing*.

Outliers can unduly influence the optimization of smoothing parameters in Exponential Smoothing models. Fortunately, these data problems are now well understood, and reasonable solutions allow forecasters to automatically pre-filter and replace unusual observations in the dataset before estimating smoothing model parameters (Gelper, Fried, and Croux 2009).

In summary, we can generalize Exponential Smoothing to many different forms of time series. A priori, it is sometimes unclear which Exponential Smoothing model to use; however, one can run a forecasting competition to determine which model works best on past data (see Chapter 18). In this context, the *innovation state space* framework for Exponential Smoothing is usually applied (Hyndman et al. 2008). This framework differentiates between five different variants of trends in time series (none, additive, additive dampened, multiplicative, multiplicative dampened) and three different variants of seasonality (none, additive, multiplicative). Further, we can conceptualize random errors as either additive or multiplicative. As a result, $5 \times 3 \times 2 = 30$ different versions of Exponential Smoothing are possible; this catalog of models is sometimes referred to as Pegels' classification. Models are referred to by three letters, the error (A or M), the trend (N, A, Ad, M, or Md, where Ad and Md stand for dampened versions), and the seasonality (N, A, or M), and each version's formulas for applying the Exponential Smoothing logic are well known (Gardner 2006; Hyndman and Athanasopoulos 2021). To make this framework work, one can define a hold-out sample, i.e., a sub-portion of the data available, and fit each of these 30 models to the data in the hold-out sample. One can then test the performance of each fitted model in the remaining data and select the one that works best. Alternatively, one can use *information criteria* in-sample to choose the best fitting model.

Using all 30 models in Pegel's classification may be excessive, and some models may be unstable, especially those combining multiple multiplicative components. A reduced set, including eight Exponential Smoothing models (ANN, ANA, AAdN, AAdA, MNN, MNM, MadN, and MAdM), is efficient and effective in large-scale tests (Fotios Petropoulos et al. 2023). For example, when compared against the complete set of 30 models in more than 50,000 time series from the M competition data, using just these eight models decreases the MASE (see Section 17.2) from 1.046 to 0.942 and cuts the computational time to about 1/3. In other words, these eight models seem to fit most forecasting situations well.

Exponential Smoothing is implemented in the `forecast` (Hyndman et al. 2023), the `smooth` (Svetunkov 2023), and the `fable` (O'Hara-Wild et al. 2020) packages for R, under the name of "ETS", which stands for error, trend and seasonality. [More precisely, what is implemented is the more abstract state space formulation discussed above.] This is a very sophisticated implementation that will automatically choose the (hopefully) best model out of the full classification of Gardner (2006), and that will automatically optimize the smoothing parameters. There are also various specific Exponential Smoothing models implemented across multiple other R packages, but we would always recommend going with `forecast` and `fable`.

In Python, Exponential Smoothing is implemented in various classes in the `statsmodels` package (Perktold et al. 2022). There is no unified framework as with ETS in the R packages, so you will need to assess each model on its merits yourself; however, the `statsmodels` classes do optimize the smoothing parameters.

Key takeaways

1. Exponential Smoothing is a simple and surprisingly effective forecasting technique. It can model trend and seasonality components.
2. Exponential Smoothing can be interpreted as feedback-response learning, as a weighted average between the most recent forecast and demand, or as a weighted average of all past demand observations with exponentially declining weights. All three interpretations are equivalent.
3. Your software will usually determine the optimal Exponential Smoothing model as well as the smoothing parameters automatically. High parameters imply that components change quickly over time; low parameters imply relatively stable components.
4. Trends will be extrapolated indefinitely. Consider dampening trends for long-range forecasts. Do not include trends in your model unless you have solid evidence that a trend exists.
5. There are up to 30 different forms of Exponential Smoothing models, although only a smaller subset of them are broadly relevant; modern software will usually select which of these works best in a time series using an estimation sample or information criteria.

10

ARIMA models

Besides Exponential Smoothing, so-called ARIMA, or *Box–Jenkins* models, represent another approach for analyzing time series. ARIMA stands for *Autoregressive Integrated Moving Average.* These models are popular, particularly among econometric forecasters. As we will see later in this chapter, ARIMA models generalize Exponential Smoothing models.

There is a logic explaining the popularity of ARIMA models: If Exponential Smoothing models are a special case of ARIMA models, then ARIMA models should outperform, or at least perform equivalently to, Exponential Smoothing. Unfortunately, though, this logic does not seem to hold in reality. In large forecasting competitions (Makridakis et al. 1993; Makridakis and Hibon 2000), Exponential Smoothing models regularly beat ARIMA models in out-of-sample comparisons (Wang et al. 2022). Nevertheless, forecasters continue to use ARIMA models, and this chapter will briefly explore the mechanics of ARIMA time series modeling.

10.1 Autoregression

The first component of an ARIMA model is the autoregressive (AR) component. Autoregression or autocorrelation (see Section 6.5) means that we conceptualize current demand as a function of previous demand. This modeling choice conceptualizes demand slightly differently (though mathematically somewhat similar) than Exponential Smoothing. If we see demand as driven by only the most recent demand, we can write the simplest form of an ARIMA model, also called an AR(1) model, as follows:

$$Demand_t = a_0 + a_1 \times Demand_{t-1} + Error_t \tag{10.1}$$

Equation (10.1) looks like a regression equation (see Chapter 11), with current demand being the dependent variable and previous demand being the independent variable. AR(1) models can be estimated this way – simply as a regression equation between previous and current demand. One can, of course, easily extend this formulation and add more lagged terms to reflect a dependency of the time series that goes deeper into the past. For example, an AR(2) model would look as follows:

$$Demand_t = a_0 + a_1 \times Demand_{t-1} + a_2 \times Demand_{t-2} + Error_t \qquad (10.2)$$

Estimating this model again follows a regression logic. Generally, an AR(p) model is a regression model that predicts current demand with the p most recent past demand observations.

Trends in a time series are usually *filtered out* before we estimate an AR model by first-differencing the data. Further, dealing with seasonality in this context of AR models is straightforward. For instance, you can deseasonalize your data before analyzing a time series by taking year-over-year differences. You can also include an appropriate demand term in the model equation. Suppose, for example, we examine a time series of monthly data with expected seasonality. We could model this seasonality by allowing the demand from 12 time periods ago to influence current demand. In the case of an AR(1) model with seasonality, we would write this specification as follows:

$$Demand_t = a_0 + a_1 \times Demand_{t-1} + a_{12} \times Demand_{t-12} + Error_t \qquad (10.3)$$

10.2 Integration

The next component of an ARIMA model is the *I*, which stands for the order of *integration*. Integration refers to taking differences of demand data before analysis, i.e., analyzing not the time series itself but analyzing a time series that has been (possibly repeatedly) differenced. For example, an AR(1)I(1) model would look as follows:

$$\begin{aligned} &Demand_t - Demand_{t-1} \\ &= a_0 + a_1 \times (Demand_{t-1} - Demand_{t-2}) + Error_t \end{aligned} \qquad (10.4)$$

To make things simple, the Greek letter Delta (Δ) is often used to indicate first differences. For example, one can write

$$\Delta Demand_t = Demand_t - Demand_{t-1} \qquad (10.5)$$

(Other symbols often seen to indicate differencing are the nabla ∇, or the *backshift operator* B, or even a D for "differencing".)

Substituting Equation (10.5) into Equation (10.4) then leads to the following simplified form:

$$\Delta Demand_t = a_0 + a_1 \times \Delta Demand_{t-1} + Error_t \qquad (10.6)$$

Taking first differences means that, instead of examining demand directly, one analyzes *changes in demand.* As discussed in Chapter 5, this technique may make the time series stationary before running a statistical model. (Note, though, that differencing can only remove one particular kind of non-stationarity, namely, trends.) We can also extend the idea of differencing. For example, an AR(1)I(2) model uses second differences by analyzing

$$\Delta^2 Demand_t = \Delta Demand_t - \Delta Demand_{t-1} \tag{10.7}$$

which is akin to analyzing the changes in the change in demand instead of the demand itself.

To see how a trended time series becomes stationary through integration, consider the following example of a simple time series with a trend:

$$Demand_t = a_0 + a_1 \times t + Error_t \tag{10.8}$$

This time series is not stationary since the mean of the series constantly changes by a factor a_1 in each successive period. If, however, we examine the first difference of the time series instead, we observe that in this case:

$$\begin{aligned} &\Delta Demand_t \\ &= (a_0 - a_0) + (a_1 \times t - a_1 \times (t-1)) + (Error_t - Error_{t-1}) \\ &= a_1 + Error_t - Error_{t-1} \end{aligned} \tag{10.9}$$

The two error terms in Equation (10.9) are simply the difference between two random variables, which is a random variable in itself. In other words, first differencing turned the time series from a non-stationary trended series into a stationary one with a constant mean of a_1. We illustrate this effect in Figure 10.1. The left part of the figure shows a non-stationary time series with a positive trend. The right-hand side of the figure shows the first differences of the same demand observations. These first differences now represent noise around a mean, thus making the series of first differences stationary.

Sometimes, first-order differencing is insufficient to make a time series stationary, and second- or third-order differencing is needed. Further, some time series require taking the natural logarithm first (which can lead to a more constant variance) or deseasonalizing the series. In practice, forecasters use many manipulations to achieve stationarity of the series, but differencing represents a widespread transformation to achieve this objective.

One can, of course, argue that all these data manipulations distract from the actual objective. In the end, forecasting is about predicting the next demand in a series and not about predicting the next first difference in demand. But notice that we can easily ex-post reverse data manipulations such as first differencing. Suppose you have used an AR(p)I(1) model to predict the next

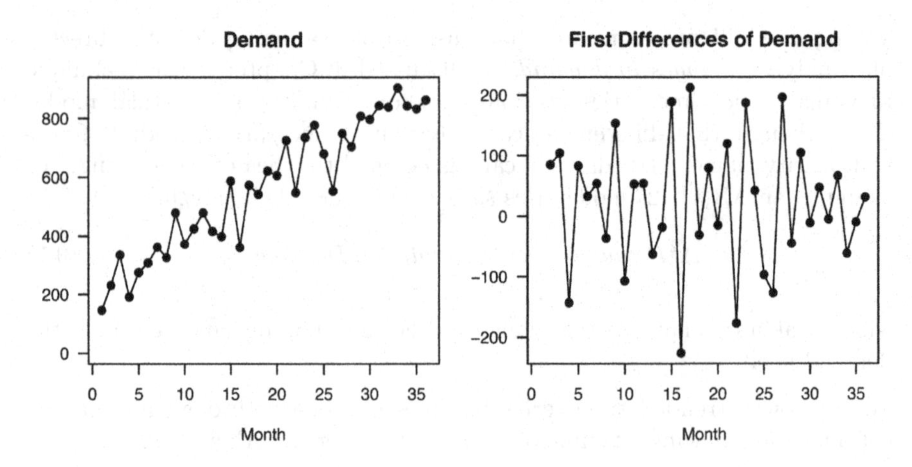

FIGURE 10.1
First differencing a time series – note the different vertical axes

first difference in demand (= $\textit{Predicted}\ (\Delta \textit{Demand})_{t+1}$). Since you know the currently observed demand, you can construct a forecast for demand in period $t + 1$ by calculating:

$$\textit{Forecast}_{t+1} = \textit{Predicted}\ (\Delta \textit{Demand})_{t+1} + \textit{Demand}_t \tag{10.10}$$

10.3 Moving averages

Having discussed the *I* component of ARIMA models, what remains is to examine the *moving averages* (MA) component. This MA component in ARIMA models should not be confused with the moving averages forecasting method discussed in Section 9.1! This component represents an alternative conceptualization of serial dependence in a time series – but this time, the subsequent demand does not depend on the previous demand but on the previous error, that is, the difference between what the model would have predicted and what we have observed. We can represent an MA(1) model as follows:

$$\textit{Demand}_t = a_0 + a_1 \times \textit{Error}_{t-1} + \textit{Error}_t \tag{10.11}$$

In other words, instead of seeing current demand as a function of previous demand, we conceptualize demand as a function of previous forecast errors. The difference between autoregression and moving averages, as we shall see later in this chapter, essentially boils down to a difference in how persistent random shocks are to the series. Unexpected shocks tend to linger long in AR models but more quickly disappear in MA models. MA models, however, are more difficult to estimate than AR models. We estimate an AR model like any standard regression model. However, MA models have a chicken-or-egg problem: one has to create an initial error term to estimate the model, and all future error terms will directly depend on what that initial error term is. For that reason, MA models require estimation with more complex maximum likelihood procedures instead of the regular regression we can use for AR models.

MA models extend similarly to AR models do. An MA(2) model would see current demand as a function of the past two model errors, and an MA(q) model sees demand as a function of the past q model errors.

More generally, when combining these components, one can see an ARIMA(p, d, q) model as a model that looks at demand that has been differenced d times, where this dth demand difference is seen as a function of the previous p (dth) demand differences and q forecast errors. Quite obviously, this is a very general model for demand forecasting, and selecting the right model among this infinite set of possible models becomes a key challenge. One could apply a "brute force" technique, as in Exponential Smoothing, and examine which model among a wide range of choices fits best in an estimation sample. Yet, unlike in Exponential Smoothing, where the number of possible models is limited, there is a nearly unlimited number of models available here since one could always go further into the past to extend the model (i.e., increase the orders p, d and q).

10.4 Autocorrelation and partial autocorrelation

Important tools forecasters use to analyze a time series, especially in ARIMA modeling, are the *autocorrelation function (ACF)* and the *partial autocorrelation function (PACF)*, as well as *information criteria*, which we consider in Section 10.5.

For the ACF, one calculates the sample correlation (e.g., using the `=CORREL` function in Microsoft Excel) between the current demand and the previous demand, between the current demand and the demand before the previous demand, and so on. Going up to n time periods into the past leads to n correlation coefficients between current demand and the demand lagged by up to n time periods (see Section 6.5). Autocorrelations are usually plotted against lag order in a barplot, the so-called *autocorrelation plot*. Figure 10.2 contains an example of such a plot. An added feature in an autocorrelation plot is a horizontal bar, which differentiates autocorrelations that are statistically significant from non-significant ones, that is, autocorrelation estimates that are higher (or lower, in the case of negative autocorrelations) than we would expect by chance. This significance test can sometimes help identify an ARIMA model's orders p and q.

The PACF works similarly. We estimate first an AR(1) model, then an AR(2) model, and so on, and always record the regression coefficients of the last term we add to the equation. These *partial autocorrelations* are again plotted against the lags in a barplot, yielding the *partial autocorrelation plot.*

To illustrate how to calculate the ACF and the PACF, consider the following example. Suppose we have a demand time series and calculate the correlation coefficient between demand in a current period and the period before (obtaining $r_1 = 0.84$), as well as the correlation coefficient between demand in a current period and demand in the period before the previous one (obtaining $r_2 = 0.76$). The first two entries of the ACF are then 0.84 and 0.76. Suppose now that we additionally estimate two regression equations:

$$Demand_t = Constant_1 + \theta_{1,1} \times Demand_{t-1} + Error_{1,t} \tag{10.12}$$

and

$$\begin{aligned} &Demand_t \\ &= Constant_2 + \theta_{2,1} \times Demand_{t-1} + \theta_{2,2} \times Demand_{t-2} + Error_{2,t}. \end{aligned} \tag{10.13}$$

Suppose that the results from this estimation show that $\theta_{1,1} = 0.87$ and $\theta_{2,2} = 0.22$. The first two entries of the PACF are then 0.87 and 0.22.

We can use the ACF to differentiate MA(1) and AR(1) processes from each other and differentiate both of these processes from demand, representing simple draws from a distribution without any serial dependence. To illustrate this selection process, Figure 10.2 contains the time series and ACF plots of a series following an AR(1) process, an MA(1) process, and data that represents simple draws from a normal distribution. We generated all three series using the same random draws to enhance comparability. You can see that the MA(1) series and the normal series look very similar. Only a comparison of their ACFs reveals that while the normally distributed demand shows no autocorrelation in any time lag, the MA(1) series shows an autocorrelation going back one time period (but not further). The AR(1) process, however, looks very distinct. Shocks from the series tend to throw the series off from the long-run average for more extended periods. In other words, if demand drops, it will likely stay below average for some time before it recovers. This distinctive pattern is visible in the ACF for the AR(1) process; the autocorrelation coefficients in the series are present in all four time lags depicted, though they slowly decrease as the time lag increases.

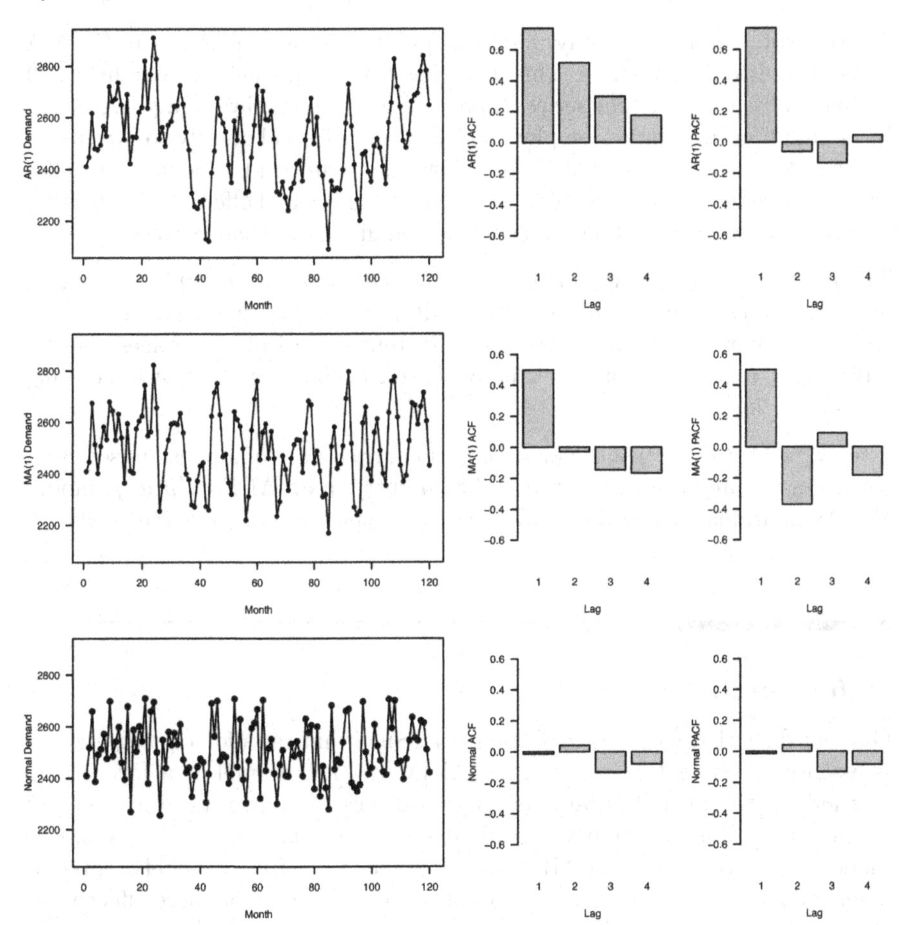

FIGURE 10.2
Autoregressive (AR), Moving Average (MA) and white noise series with their ACF and PACF

10.5 Information criteria

As discussed above, we can use (P)ACF plots to identify the orders of pure AR(p) and pure MA(q) processes and to distinguish these from each other. However, (P)ACF plots are not helpful for identifying general ARIMA(p, d, q) processes with both $p, q > 0$ or $d > 0$. Since, in practice, we never know beforehand whether we are dealing with a pure AR(p) or MA(q) process, the use of (P)ACF plots in order selection is limited (Kolassa 2022b). For order selection in this more general case, one uses *information criteria.*

Information criteria typically assign a quality measurement to an ARIMA model fitted to a time series. This value has two components. One is how well the model fits the past time series. However, since we can always improve model fit by simply using a more complex model (which will lead quickly to overfitting and bad forecasts, see Section 11.7), information criteria also include a penalty term that reduces the value for more complex models. Different information criteria differ mainly in how exactly this penalty term is calculated.

The most used information criteria are *Akaike's information criterion* (AIC) and the *Bayes information criterion* (BIC). These two information criteria (and other more rarely used ones) have different statistical advantages and disadvantages, and neither one is "usually" better in choosing the best forecasting model.

Finally, neither the (P)ACF nor information criteria are of any use in selecting the order of integration, i.e., the middle d entry in an ARIMA(p, d, q) model. For this, automatic ARIMA model selection typically uses *unit root tests.*

10.6 Discussion

One can show that using single Exponential Smoothing as a forecasting method is essentially equivalent to using an ARIMA$(0, 1, 1)$ model as a forecasting method. With optimal parameters, the two series of forecasts produced will be the same. The apparently logical conclusion is that since ARIMA$(0, 1, 1)$ models are a special case of ARIMA(p, d, q) models, ARIMA models represent a generalization of Exponential Smoothing and thus must be more effective at forecasting.

While this logic is compelling, it has not withstood empirical tests. Large forecasting competitions have repeatedly demonstrated that Exponential Smoothing models tend to dominate ARIMA models in out-of-sample comparisons (Makridakis and Hibon 2000; Makridakis, Spiliotis, and Assimakopoulos 2022), although ARIMA has performed well in quantile forecasting in the M5 competition (Makridakis, Petropoulos, and Spiliotis 2022). It seems that Exponential Smoothing is more robust. Further, with the more general state space modeling framework (Hyndman et al. 2008), the variety of Exponential Smoothing models has been extended such that many of these models are not simply special cases of ARIMA modeling anymore. Thus, ARIMA modeling is not necessarily preferred as a forecasting method in general; nevertheless, these models continue to enjoy some popularity.

ARIMA modeling is a recommended forecasting approach for a series showing short-term correlation and not dominated by trend and seasonality. However, forecasters using the method require technical expertise to understand how to carry out the method (Chatfield 2007).

In using ARIMA models, one usually does not need to consider a high differencing order. We can model most series with no more than two differences (i.e., $d \leq 2$) and AR/MA terms up to order five (i.e., $p, q \leq 5$) (Ali, Boylan, and Syntetos 2012).

ARIMA models focus on point forecasts, although they can also output prediction intervals and predictive densities based on an assumption of constant-variance Gaussian errors. A similar class of models, called *Generalized Autoregressive Conditional Heteroscedasticity (GARCH)* models, focuses on modeling the uncertainty inherent in forecasts as a function of previous shocks (i.e., forecast errors). Forecasters use these models in stock market applications. They stem from the observation that large shocks to the market create more volatility in the succeeding periods. The key to these models is to view the variance of forecast errors not as fixed but as a function of previous errors, effectively creating a relationship between previous shocks to the time series and future uncertainty of forecasts. GARCH can be combined with ARIMA models since the former technique is focused on modeling a distribution's variance while the latter models the mean. An excellent introduction to GARCH modeling is given in Engle (2001) and Batchelor (2010).

ARIMA models, with automatic order selection ("auto-ARIMA") are implemented in the `forecast` (Hyndman et al. 2023) and the `fable` (O'Hara-Wild et al. 2020) packages for R. The `pmdarima` package for Python (T. G. Smith et al. 2022) ports this functionality from the R packages.

Key takeaways

1. ARIMA models explain time series based on autocorrelation, integration, and MAs. We can apply them to time series with seasonality and trend.
2. These models have not performed exceptionally well in demand forecasting competitions but are included in most software packages.
3. Your software should automatically identify the best differencing, AR, and MA orders based on unit root tests and information criteria and estimate the actual AR and MA coefficients.
4. ACF and PACF plots are often used in ARIMA modeling to investigate a time series. However, identifying the best ARIMA model based on these plots rarely works.
5. You will typically not need high AR, MA, or differencing orders for the model to work.

11

Causal models and predictors

The time series forecasting methods discussed in the previous chapters require only a time series history. No additional data are needed to calculate a simple forecast or to estimate model parameters. However, organizations often have various other pieces of information beyond the time series history. Examples may include consumer confidence indices, advertising projections, reservations, population change, weather, etc. Such data may explain variations in demand and could serve as the basis for more accurate demand forecasts. Using such information is the domain of *causal modeling* or *regression modeling*.

For instance, suppose you are interested in producing a monthly sales forecast of a particular product in a company. You want to investigate the association between advertising expenses and sales. The goal is to develop a model that you can use to forecast sales based on advertising expenses. In this setting, advertising expenses are an input variable, while sales are an output variable. The inputs go by different names, such as *predictors*, *features*, *drivers*, *indicators*, or *independent*, *exogenous* or *explanatory variables*. The output variable is often called the *response*, or the *explained*, *dependent* or *forecast variable*. Throughout this chapter, we will use the terms "forecast variable" for the output and "predictor" for the input.

11.1 Association and correlation

This section examines how to describe and model an association between two or more variables, and to leverage it in forecasting. To be more precise, we are interested in the association between the forecast variable (e.g., demand or sales) and some potential predictors (e.g., advertising expenses, price or weather). The critical question is whether knowing the values of predictors provides any information on the forecast variable. In other words, does the knowledge of a predictor's values increase forecast accuracy? For example, if weather and sales are independent, then knowing the weather forecast does not increase sales forecast accuracy. There is no value in including it in the model. But if promotions influence sales, learning about promotions in advance and incorporating this predictor into a forecasting model will increase accuracy.

A scatterplot is a convenient way to demonstrate how two numerical variables are associated. We could, in addition, quantify the association using a numerical summary. There are many ways to do so. The Pearson correlation coefficient may be the most commonly used (Benesty et al. 2009). This coefficient measures the strength and direction of a *linear* association. (Other correlation coefficients, like Spearman's or Kendall's, also measure *non-linear* – but still monotonic – correlations.) The Pearson correlation coefficient r can take any value between -1 and $+1$. A perfect correlation, i.e., $r = +1$ or $r = -1$, would require all data points to fall on a straight line. Such a correlation is rare, because of randomness.

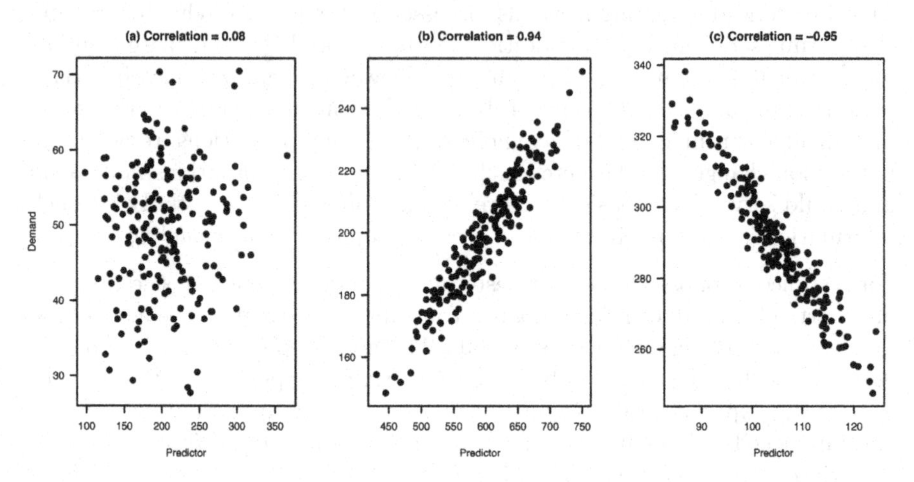

FIGURE 11.1
Correlations between two numerical variables

Intuitively, if there is no relationship between variables, we would expect to see no patterns in the scatterplot. It will be a cloud of points that appears to be randomly scattered. We observe this in Figure 11.1(a). Demand seems to have no particular pattern, regardless of the predictor's value. We would not expect including this predictor in a model to lead to more accurate demand forecasts. On the other hand, Figure 11.1(b) and (c) illustrate examples with a strong positive and negative linear association between demand and the predictor, respectively. As the predictor changes on the horizontal axis, so does demand on the vertical axis. In both cases, knowledge of the predictor value provides information about the corresponding demand, and including the predictor in a causal model will likely improve forecasts.

The terms "association" and "correlation" are sometimes used synonymously, but they do not mean the same thing. While a *correlation* is more specific, measuring a monotonic (typically linear) association between two numerical variables, *association* is a more general term and could refer to any linear or non-linear relationship. Figure 11.2 illustrates the difference. Figure 11.2(a)

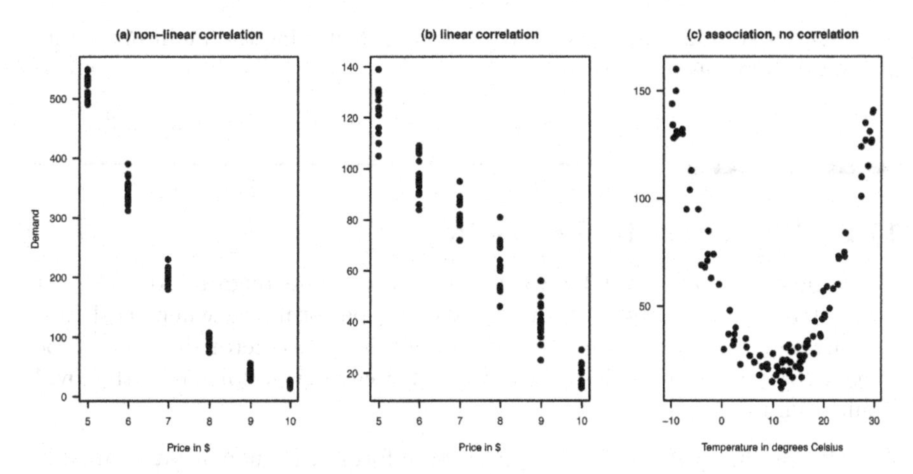

FIGURE 11.2
Association vs. correlation between two numerical variables

shows a *non-linear* correlation between price and demand: a price change from \$9.99 to \$8.99 yields a much smaller demand uplift than a price change from \$5.99 to \$4.99. Figure 11.2(b) shows a *linear* correlation, where every price reduction by \$1 yields about the same additive demand uplift. Both linear and non-linear correlations are associations. Finally, Figure 11.2(c) shows a clear association between temperature and demand, e.g., for sunscreen in a mountain resort: low temperatures happen in winter, when skiers need sunscreen, high temperatures correspond to summer, where hikers need sunscreen, and middling temperatures correspond to the off-season in spring and fall with low demand for sunscreen. Temperature and demand are uncorrelated since there is no *linear* relationship between the two, but of course, the obvious association will help us forecast better (given good weather forecasts).

Many forecasters use the correlation coefficient to check for an association between variables in a dataset. However, this measure can be misleading. Examine the scatterplot in Figure 11.2(a). The association between price and demand is not linear. However, we can still calculate a correlation coefficient. The Pearson correlation in this example is not very useful, since the underlying relationship is non-linear. Further, Figure 11.2(c) has a zero correlation between temperature and demand. This measure is again misleading, as it does not capture the clear non-linear association between the two variables.

In summary, we should only calculate the Pearson correlation when we need to measure the *linear* strength between two numerical variables. We should not use this measure when there is a *non-linear* association; otherwise, you get deceiving results. Figure 11.2 also reiterates the importance of plotting your data using a scatterplot and not exclusively relying on the correlation coefficient when trying to understand data.

An essential step in building causal models is identifying the main predictors of the variable you want to forecast.

11.2 Predictors in forecasting

Useful predictors are available (or predictable) early enough in advance, not prohibitively costly, and improve forecasting performance when used in a forecasting model. Of course, the cost of obtaining predictor data has to be judged compared to the improvement in forecasting performance that will result from using it.

A useful predictor needs to provide more information than what is already contained in the time series itself. For instance, weather data are often subject to the same seasonal effects as the time series whose forecasts we are actually interested in, e.g., the demand for ice cream. Simple correlations between a predictor and the focal series can result from their joint seasonality. In this case, simply including seasonality in the model may make modeling the predictor superfluous. A key to a successful evaluation of the performance of predictors is not merely to demonstrate a correlation or association between the predictor and your demand time series, but also to show that using the predictor in forecasting improves upon forecasts when used in addition to the time series itself.

To use a predictor for forecasting, we need to know or be able to forecast its values for the future. We may be able to predict our own company's advertising budgets rather well. Public holidays, concerts, festivals, or promotions are categorical variables known in advance. Conversely, assume that we wish to forecast the sales of a weather-sensitive product like garden furniture or ice cream for the next month. If the weather is nice and sunny, we will sell more than if it is rainy and wet. However, suppose we want to use weather information to improve forecasts. In that case, we must feed the forecasted weather into our causal forecast algorithm. Of course, the question is whether we can forecast the weather sufficiently far enough into the future to improve sales forecasts that do not use the weather – where we assume we have already included seasonality in our "weatherless" forecasts, since these products are typically strongly seasonal.

Importantly, when evaluating the performance of your forecast that leverages weather data, you need to ensure that you don't assess your out-of-sample forecasts based on how they work with *actual* weather. You will not know next week's actual weather when you produce the forecast for the next week. You need to assess the forecasts based on weather predictions. The uncertainty in these weather predictions adds to the uncertainty in your forecasting model, which ultimately affects the forecast accuracy (Satchell and Hwang 2016). It is thus very much recommended to begin the search for useful predictors with those whose values we know in advance (*deterministic* predictors), rather than those that require forecasting themselves (*stochastic* predictors).

Sometimes the association between a predictor and an outcome is not instantaneous. In the following sections, we will discuss leading and lagging predictors as two important sources of information.

11.3 Leading and lagging predictors

A *leading predictor* (especially in macroeconomics often called a *leading indicator*) is a numerical or categorical predictor time series containing predictive information that can help increase forecast accuracy for a different forecast variable at a *later* point in time (the predictor "leads" the focal time series). Examples would be housing starts as a leading predictor for roof construction, higher inflation today leading to less economic activity later, a higher volume of calls received in a clinical desk service leading to higher demand for emergency room service a few hours later, or a promotion on a product today leading to lower demand next week, because customers have stocked up on the product ("pantry loading").

In contrast, there are also *lagging* or *lagged predictors* (again, often called *lagged indicators*), whose effect *follows*, or *lags behind* the predictor. For example, if customers know that a product will be on promotion at a lower price point next week, they may refrain from buying it today, so today's demand is lower because of next week's promotion. And known regulatory or tax law changes that will take effect months or years from now may have an impact on economic activity today: fewer companies will invest in building new gas stations if sales of internal combustion engine cars will be illegal two years from now.

We see that predictors can have an impact over multiple time periods: a promotion may lead to higher demand *during* the promotion, but lower demand *after* the promotion (because people have stocked up) or even *before* the promotion (if customers can predict the promotion, e.g., if promotions on a given product happen regularly or are communicated in advance). Figure 11.3 illustrates the effect of a leading or lagging predictor.

11.4 An example: advertising and sales

To illustrate the use of predictors, consider a dataset on sales and advertising. The dataset contains monthly sales and advertising expenses for a product

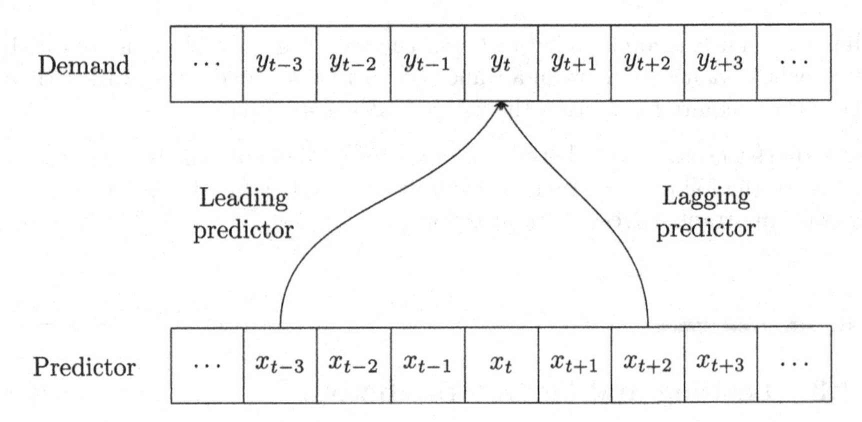

FIGURE 11.3
A leading or lagging predictor for demand

between January 2005 and August 2021 (Kaggle 2023). One could apply some form of Exponential Smoothing to the sales data to create a forecast – or one could attempt to use advertising expenses as a predictor for sales.

An easy way to examine whether advertising expenses are associated with higher sales is to calculate the correlation coefficient between these variables during the same period. The left panel of Figure 11.4 gives a scatterplot of the dataset with its correlation coefficient.

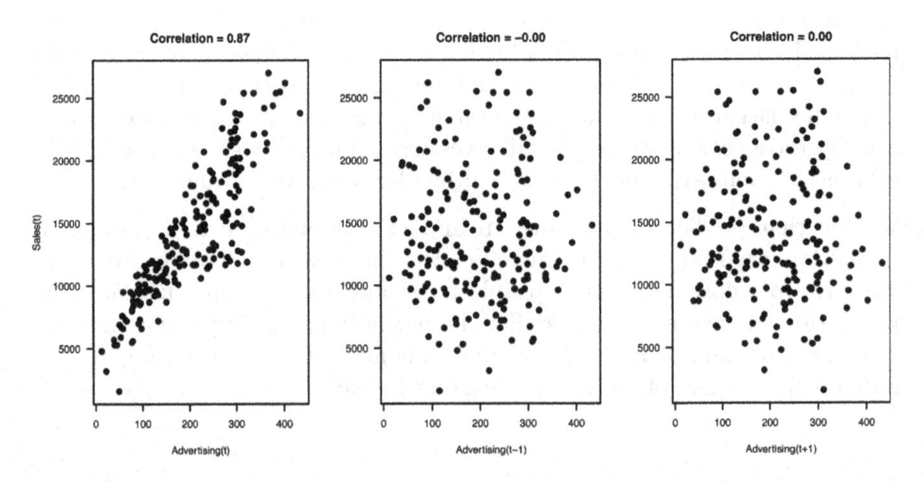

FIGURE 11.4
The association between time series on sales and advertising (left), as well as the association between sales and lagging advertising (center) and leading advertising (right)

So is advertising a useful predictor? The first relevant question would be whether we know a month's advertising expenses in advance. Usually, firms set advertising budgets according to some plan so that many firms will know their advertising expenses for a future month in advance. However, the time lag with which this information is available effectively determines the possible forecast horizon. If we know advertising expenses only one month in advance, we can use this information for one-month-ahead predictions only. As highlighted in Figure 11.4, the association between sales and advertising is strong and linear. Therefore, advertising should be considered as an essential predictor of sales when building the forecasting model.

Revisiting Figure 11.4, we can investigate whether there is a time lag between spending money on advertising and these expenses having any influence on sales. Advertising is sometimes seen as filling the front end of a funnel, starting the customers on their journey. Using lag plots (see Figure 6.7), one can check for a possible time lag by visualizing the association between sales and *previous* or *later* advertising expenses. Additionally, we can compare the correlation coefficients from the current period ($r = 0.87$ for the example data above) to the correlation between sales and last month's advertising expenses ($r = 0.00$), or even the expenses from one month later ($r = 0.00$). The correlation is strongest between the current advertising expenses and sales, indicating that a time lag between these variables is not essential for forecasting in this example. If we had found a stronger relationship between sales and leading or lagging advertising, we could have continued our analysis by considering longer lag or lead times.

11.5 Correlation, causation and forecasting

While we use the term "causal models" in the title of this chapter, truly establishing causality is very different from showing a correlation or estimating a regression equation (Pearl and Mackenzie 2018). Doing so requires an association between one variable and another, temporal precedence, and the exclusion of alternative explanations. In our example of advertising and sales, one could argue for *reverse causality* (i.e., it is not advertising that is driving sales, but sales that is driving advertising, because higher sales may increase the budget available for advertising) or alternative explanations (i.e., if companies advertise, they also stock more product, leading to higher service levels and sales).

The gold standard in establishing causality is the randomized controlled trial. One assigns a *treatment*, i.e., a manipulation of an input variable, at random to one of two similar groups – the treatment and the control groups. One can then establish causality by comparing both groups' outcome measurements afterward. While going through this process is often tricky in practice (running a promotion in only half of stores risks alienating customers), the methodology of improving websites through A/B testing is similar to a randomized trial.

For non-experimental data, causality is empirically hard to establish. However, the field of econometrics has made much progress in recent years to examine causality better. We refer interested readers to Angrist and Pischke (2009) for a good overview of these methods. Examples of these methods include difference-in-difference analysis, where we compare a treated group over time to a similar non-treated group. Or regression-discontinuity approaches, where we examine "close calls" in our data. For instance, suppose we want to examine the effect of unionization on outcomes. In that case, we focus our analysis on only the sites where the vote to unionize was very close and could have gone either way. Whether the site ended up unionized or not is thus somewhat randomly determined. Comparing those "close call" sites that unionized to those that did not is thus almost like comparing treated and control groups in a randomized controlled trial.

Establishing a causal association helps us understand why a predictor explains variation in the forecast variable. This understanding can increase trust in the model and lead to a higher chance of implementation and use in practice. Without establishing causality, an observed association or correlation could be *spurious*, i.e., not driven by a "true" causal link, but merely due to random noise.

We emphasize that forecasting does not *require* a causal relationship; in this sense, it is a very pragmatic profession. As long as data enable us to predict the future better, we do not need to be sure that the underlying relationship is genuinely causal in the sense of, e.g., Pearl and Mackenzie (2018). While we can have more confidence that the relationship we use for forecasting remains stable over time if we understand the underlying causality, being unable to demonstrate causality does not necessarily prevent us from exploiting an empirical relationship to make predictions. If we know that a statistical relationship exists between the number of storks in a country and its birth rate ($r = 0.62$; see Matthews 2000), can we use this relationship to predict the birth rate of a country for which we only know the number of storks fluttering about? The answer is yes. Exploiting this relationship may be the best we can do without other data. However, understanding causality may lead us to better predictors. In this case, the number of storks in a country relates to country size, which relates to birth rates. Using country size will probably lead to better predictions of births than using the number of storks.

In this context, *big data* provides forecasters with new opportunities to find predictors for their time series. Google, for example, provides a tool available for free: Google Trends (Choi and Varian 2012). Google Trends allows forecasters to examine the frequency of specific search terms relative to all searches over time. An example focusing on end-consumer personal consumption expenses in different categories shows how incorporating this information into standard forecasting methods can increase forecasting accuracy (Schmidt and Vosen 2013). Google Trends also allows downloading and using this data for real-time forecasting.

Big data could mean more predictors and potentially more helpful associations. However, if there are too many predictors but too few observations per predictor, then spurious relationships are more and more likely to appear. In other words, big data may mean more information but also more false information (Taleb 2014, see also Section 11.7 below).

11.6 Combination with time series

Let us return to our advertising and sales data from Section 11.4. The critical question is how much we gain by using advertising data in our forecast. That is, how much better we can predict sales if we exploit the relationship between advertising and sales, as opposed to simply using the time series history of sales to predict the future?

To answer this question, we devise a simple forecasting competition (see Chapter 18) by splitting the dataset at hand into an estimation sample (January 2005 to August 2020) and a hold-out sample (September 2020 to August 2021). We use the estimation sample to estimate the relationship between advertising and sales; we use the hold-out sample to test the predictions of that model. Specifically, we will calculate a *rolling origin forecast*. That is, we take the history from January 2005 to August 2020, fit a model, and forecast for September 2020. Next, we move the forecast origin by one month, taking the history from January 2005 to September 2020, fitting a model and forecasting for October 2020, and so forth. In each step, we estimate the following regression equation using data from the current history:

$$Sales_t = a_0 + a_1 \times Advertising_t + Error_t, \tag{11.1}$$

Estimating a regression equation here means finding values of a_0 and a_1 that minimize the squared values of $Error_t$ across the estimation sample. We refer interested readers who require more background on regression to Richardson (2011) and Harrell (2015). We can estimate a regression equation in Microsoft Excel if we install the Analysis ToolPak or similar add-ons. In our case, in the first step, we estimate the equation to obtain the following estimates for a_0 and a_1:

$$Sales_t = 4,241.03 + 48.77 \times Advertising_t + Error_t \tag{11.2}$$

These estimates indicate that every dollar spent in advertising is associated with $48.77 in extra sales for prediction purposes on average. We emphasize that estimating this regression does not establish a *causal* effect, and we should not necessarily use these numbers to plan advertising spending. However, if we know the advertising budget, we can multiply that number by 48.77 (and add 4,241.03) to predict sales in that period.

In the second step, we have one additional historical data point, so our estimated model changes slightly:

$$Sales_t = 4,177.95 + 48.74 \times Advertising_t + Error_t \tag{11.3}$$

This rolling regression approach is easy to do for all 12 months in the hold-out sample (i.e., last year) and leads to a Mean Absolute Error (MAE; see Chapter 17 on forecast error measures) of 1,954 across the hold-out sample.

How good is this forecast? A reasonable comparison is a pure time series forecast that would not have required any information on advertising. To that purpose, we again used rolling forecasts to estimate Exponential Smoothing forecasts (as described in Chapter 9). The winning models across our 12 forecast origins generally were models with additive or multiplicative errors, without trends or seasonality. The resulting MAE from these model forecasts is 4,810, much higher than the MAE from our simple regression model.

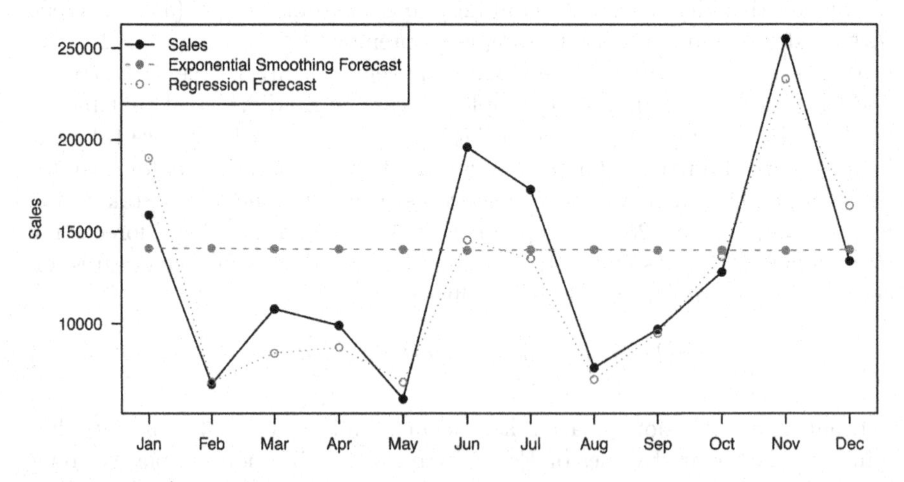

FIGURE 11.5
Rolling regression and Exponential Smoothing forecasts in the hold-out sample

We can see from Figure 11.5 that the two sets of forecasts are very different. Which forecast should one trust? The MAE from Exponential Smoothing models is higher. But does that mean that one should rely solely on advertising data? Or can these two methods somehow be combined to provide better forecasts?

One way to combine time series models with predictors is to allow the regression equation to estimate seasonal factors (and trends). To incorporate seasonality into our regression equation, we code 11 *dummy* variables in our dataset, one for each month except December (see Section 15.5). A code of 1 indicates that a particular observation in the dataset takes place in that specific month; the variables are 0 otherwise. If all variables are coded as 0 for an observation, that observation took place in December. We then estimate the following multivariate regression equation:

$$\begin{aligned} Sales_t = a_0 + a_1 \times Advertising_t + a_2 \times January + \cdots + \\ a_{12} \times November + Error_t \end{aligned} \tag{11.4}$$

This equation now accounts for seasonality according to different months. However, seasonality is additive in the model and cannot change over time. We calculate forecasts for the hold-out samples using this revised specification. The resulting MAE of 2,590 is worse than in the original regression model, indicating that incorporating seasonality in this fashion was not beneficial.

This question of forecasting with multiple methods has been studied extensively in the literature on forecast combination (compare Section 8.6). In our case, we could take the average of the Exponential Smoothing and the regression forecast. While such a strategy often works to improve performance (see Section 8.6), in our case, the resulting MAE is not improved, but increases to 2,861.

Finally, while we used a straightforward multiple linear regression model in this example, the same approach can be used instead with more modern Artificial Intelligence and Machine Learning tools. Most, if not all, of these can leverage predictors and seasonal dummies for forecasting. See Chapter 14 for more information on these methods.

11.7 Model complexity and overfitting

Our forecasts are never as accurate as we want them to be. One widespread reaction to this state of affairs is to search for more and more predictors and include them in a causal model. Unfortunately, there are two different problems with this approach.

The first problem is that the proposed predictor may only spuriously correlate with our outcome. If we mine all the available data, random noise means that there will *always* be some other time series that correlates well historically with our outcome. However, this correlation may be wholly spurious and useless in forecasting. (An internet search for "spurious correlations" yields many entertaining examples.) Unfortunately, humans are highly prone to seeing patterns where none exist (an effect known as *pareidolia*) and invent

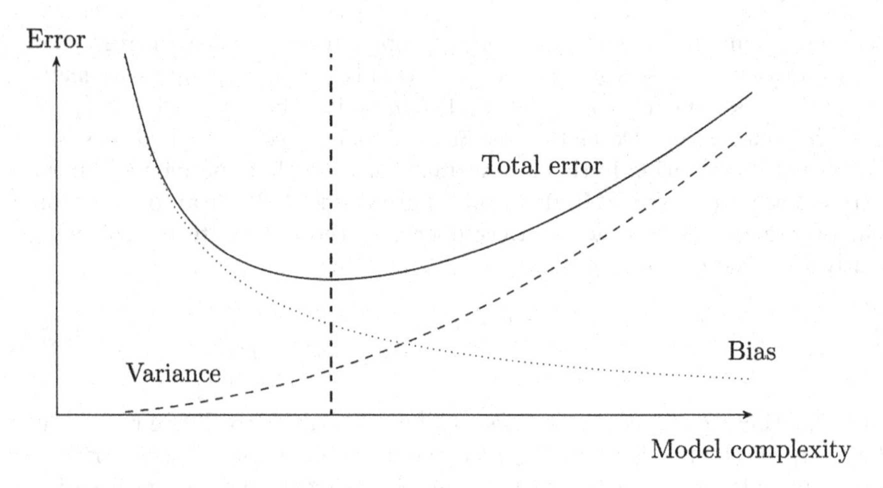

FIGURE 11.6
Model complexity and the bias-variance trade-off. The optimal model complexity – with the lowest total error – is indicated

explanations why a given predictor is "really" related to the outcome of interest and so "should" have predictive power – even if it objectively doesn't.

The problem gets even worse. Based on the preceding paragraph, we might think that including irrelevant predictors may not be helpful, but at least it will not harm us. Right? Sadly, this comforting thought is mistaken: adding spurious predictors to a model may actively make the model and the forecast *worse*. Why? Adding one predictor to a model changes the *entire* model. The estimated influence of all other predictors changes in light of the newly added one. If an additional predictor is only spuriously related to the outcome, including it in our model only adds noise. Since the entire model changes, this added noise makes the estimates for the impact of other predictors noisier. The result will be a more noisy (i.e., less accurate) forecast.

Another problem occurs when a predictor is indeed related to the outcome, but only weakly so. Including it in our model could lead to more accurate forecasts on average. However, the predictor will also make the entire model more noisy. Whether the improvement in performance by adding a variable is greater than the deterioration in quality caused by the higher model noise is uncertain. If a predictor's effect is particularly weak, the added model noise by including this predictor will likely offset any accuracy improvement, making forecasts worse overall (Kolassa 2016b).

This problem is a direct consequence of the *bias-variance trade-off* (James et al. 2021), and an example of *overfitting.* Every Data Scientist and forecaster should be aware of these concepts. Figure 11.6 shows the bias-variance trade-off and its relation to model complexity. If the model is simple and understandable, it will have a high bias and a low variance. Similarly, if the model is complex and challenging to understand, it will have a high variance and a low bias. There is a clear trade-off: as one increases, the other will decrease and vice versa. What is most important is the total error, which is the sum of the bias and the variance (the *bias-variance decomposition* of total error).

Another relevant issue is the trade-off between model complexity and model interpretability. Figure 11.7 illustrates this trade-off for some of the methods that we cover in this book in Chapters 8 to 15. In general, as the complexity of a method increases, its interpretability decreases. However, it is often the case that multiple quite different models yield very similar forecasts (such a set of models has been called a *Rashomon set* by Fisher, Rudin, and Dominici 2019, after a 1950 movie directed by Akira Kurosawa, where multiple characters describe the very same event in wildly different ways). The critical question then is not to find the most accurate, but the overall best model within this set, in terms of accuracy, interpretability and other qualities like low runtime or memory requirements.

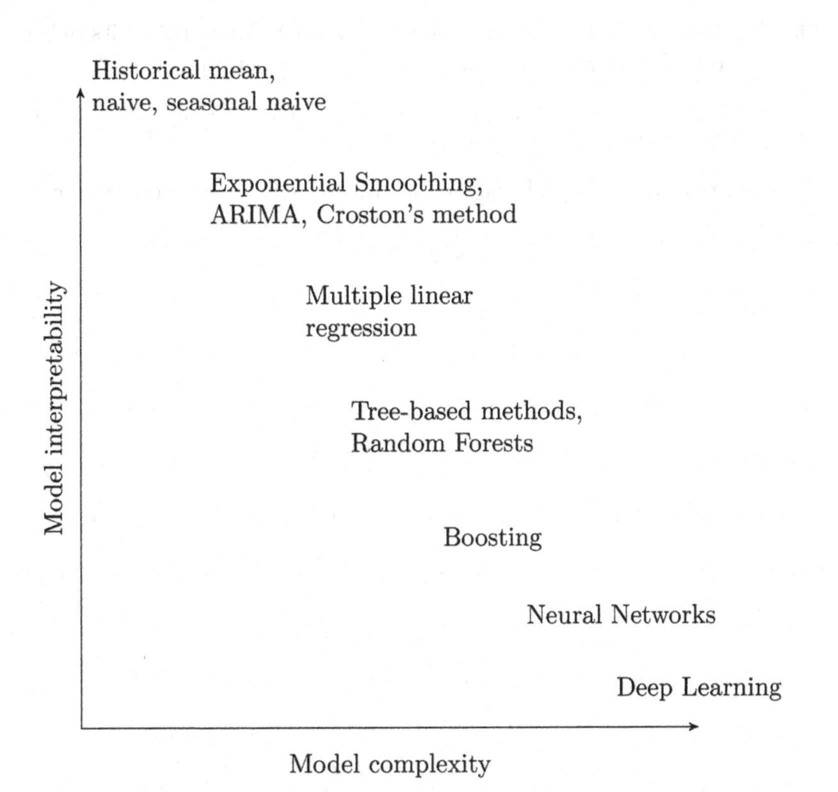

FIGURE 11.7
Model complexity and interpretability

Finally, it is essential to note that the effects described here do not only happen with *causal* predictors in the strict sense of the word. They can equally likely occur with trend and seasonality. Forecasting a weakly seasonal time series with a seasonal model may yield worse forecasts than with a non-seasonal model. Kolassa (2016b) gives a simulation example of a set of time series whose seasonality is evident in the aggregate but where a seasonal model yields worse forecasts than a non-seasonal model when applied to each constituent time series separately.

Key takeaways

1. Known predictors of demand can dramatically improve your forecast. Consider including them in a causal model.
2. It is not sufficient to demonstrate that a driver correlates with demand; rather, the cost of obtaining it needs to be outweighed by the forecast accuracy improvements it can bring.
3. A predictor's influence on demand may lag behind or lead its occurrence and may influence more than a single period.
4. For forecasting demands with a causal predictor, we need to measure or forecast the driver.
5. More complex models do not automatically yield more accurate forecasts.

12

Count data and intermittent demand

The forecasting methods we have introduced so far implicitly assumed *continuous* demand and used the normal distribution (see Figure 3.1) "under the hood." This is a reasonable approximation for fast moving products. In this chapter, we discuss specialized forecasting methods that are better suited to dealing with slow moving products.

12.1 Definitions

So far, we have focused on using the normal distribution for forecasting. Using the normal distribution can be inappropriate for two reasons: First, the normal distribution is *continuous*, i.e., a normally distributed random variable can take non-integer values, like 2.43. Second, the normal distribution is *unbounded*, i.e., a normally distributed random variable can take negative and positive values. Both these properties of the normal distribution do not make sense for (most) demand time series. Demand is usually integer-valued, apart from products sold by weight or volume, and demand is usually zero or positive, but not negative, apart from returns.

These seem obvious ways in which the normal distribution deviates from reality. So why do we nevertheless use this distribution? The answer is simple and pragmatic: *because it works*. On the one hand, using the normal distribution makes the statistical calculations that go on "under the hood" of your statistical software (optimizing smoothing parameters, estimating ARIMA coefficients, calculating prediction distributions, etc.) much more manageable from a mathematical point of view. On the other hand, the actual difference between the forecasts – point, interval and distribution – under a normal or a more appropriate distribution is often tiny, especially for fast-moving products. However, this volume argument supporting the normal distribution does not hold for slow-moving demand time series.

We can address both above-mentioned problems by using *count data distributions*, i.e., random number distributions that only yield integer values. Common distributions to model demands are the *Poisson* and the *negative binomial* distribution (Boylan and Syntetos 2021; Syntetos et al. 2011). The Poisson distribution has a single parameter representing its mean and variance. The more flexible negative binomial distribution has one parameter for the mean and another one for its variance or its over-dispersion (i.e., the amount by which the variance of the distribution exceeds its mean).

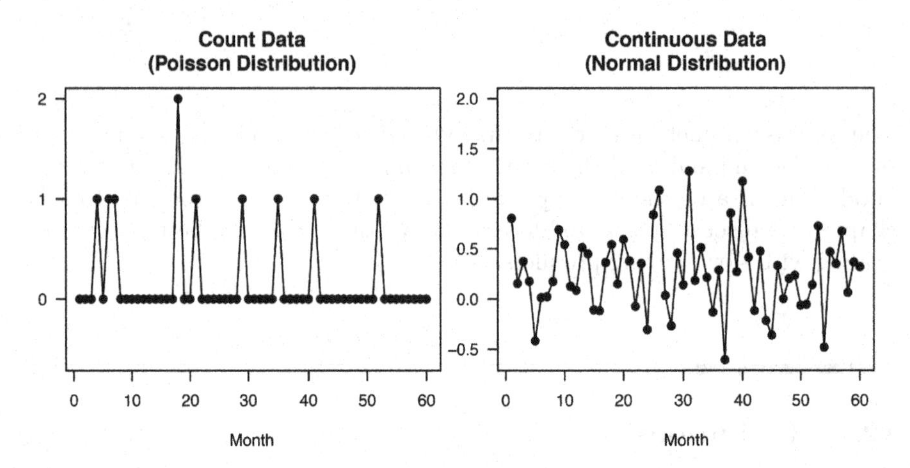

FIGURE 12.1
Random draws for count data and continuous data for low-volume products with mean and variance both equal to 0.2. The right-hand panel obviously cannot correspond to demand of a real product

Figure 12.1 illustrates the difference between Poisson-distributed and normally distributed demand at a rate of 0.2 units per month. We see both problems discussed above (negative and non-integer demands) in the normally distributed data, whereas the Poisson time series does not exhibit such issues and therefore appears more realistic. Conversely, Figure 12.2 shows little difference between a Poisson and a normal distribution for fast-moving products – here, at a rate of 20 units per month. Thus, we can reasonably model fast moving products (as in Figure 12.2) using a normal distribution, but not slow moving products (as in Figure 12.1).

Once demand gets so slow that buckets exhibit zero demand many times, we speak of *intermittent* demand series. We will interchangeably use the terms *count data* and *intermittent demand.*

Finally, there is *lumpy demand.* Demand is "lumpy" if it is intermittent *and* non-zero demands are high. Lumpy demand can occur, for instance, for upstream supply chain members with only a few customers that place batch orders. Another example is home improvement retail or wholesale stores, where builders

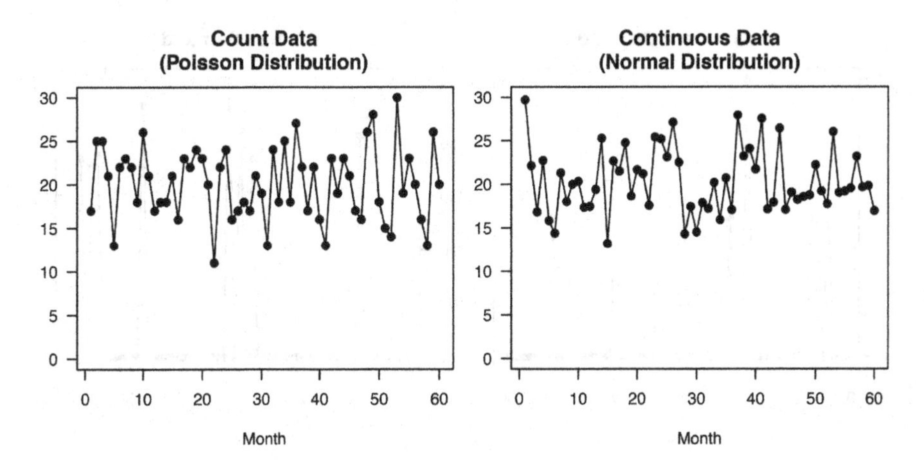

FIGURE 12.2
Random draws for count data and continuous data for high-volume products with mean and variance both equal to 20

typically buy large quantities of a particular paving stone or light switch at once. Low-volume intermittent data often results from many customers placing orders rarely; high-volume lumpy demand results from a few customers aggregating their orders into large batches. We illustrate the differences and similarities between these concepts in Figure 12.3.

Why is it important to forecast intermittent or lumpy demand? We naturally focus on forecasting the fastest-moving products simply because these products have the highest visibility in the firm (and market) and are often the most important in terms of margin and total revenue. However, most businesses also have a *Long Tail* of slow-moving products. By the anecdotal Pareto principle, 80% of your SKUs will be responsible for 20% of your sales. Classic A-B-C analysis in inventory management naturally differentiates between these fast- and slow-moving items. Many of these 80% of SKUs probably have intermittent demand series. While improving the forecasts for the 20% of fast-movers that drive 80% of sales is crucial, the many more slow-movers may represent a much more significant fraction of your total inventory value. Accurate forecasts can help you reduce these inventories, pool them, move to a make-to-order process, or generally improve your operations. There may be as significant an improvement opportunity here as there is for faster-moving items.

In addition, intermittent time series occur more and more frequently due to several recent developments. In the past, database capacity and processing power limited the number of demand time series to be stored and forecasted weekly. Nowadays, vastly more powerful storage and processing (Januschowski et al. 2013) allow working with ever-lower granularity. And a time series that is fast-moving on a weekly basis may well be slow-moving on a daily basis and

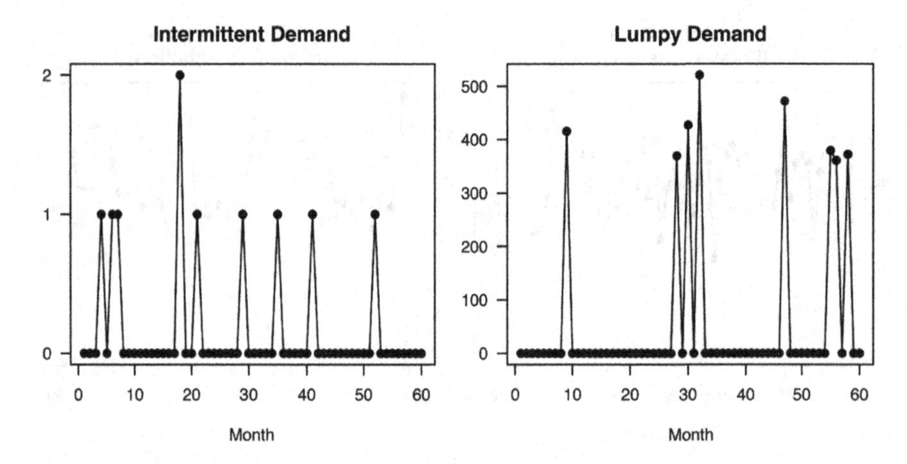

FIGURE 12.3
Intermittent and lumpy demand series. Note vertical axes.

can be heavily intermittent on an hourly basis. The more we disaggregate the time unit used for forecasting, the more likely we are to encounter intermittent data.

Further, a small product portfolio that slices a market into only a few segments by only providing a small number of product variants will likely produce high-volume series. However, with more and more product differentiation, many variants may become intermittent in demand. Thus, due to increased product variety and data storage capacity available, we need to forecast more and more intermittent time series.

12.2 Traditional forecasting methods

Given the particularities of intermittent demand, how should we forecast series with such count data? Could we also use the methods described in the previous chapters in this context?

Let us apply Single Exponential Smoothing (see Chapter 9) with a smoothing parameter of $\alpha = 0.10$ to an intermittent demand series in Figure 12.4. Remember that Single Exponential Smoothing creates forecasts by calculating a weighted average between the most recent forecast and the most recent demand. Thus, the forecasts tend to slowly move toward zero when we observe no demand. After we observe some demand, forecasts briefly *jump up* again. Thus, our forecast is high after a non-zero demand and low after a long string of zero demands.

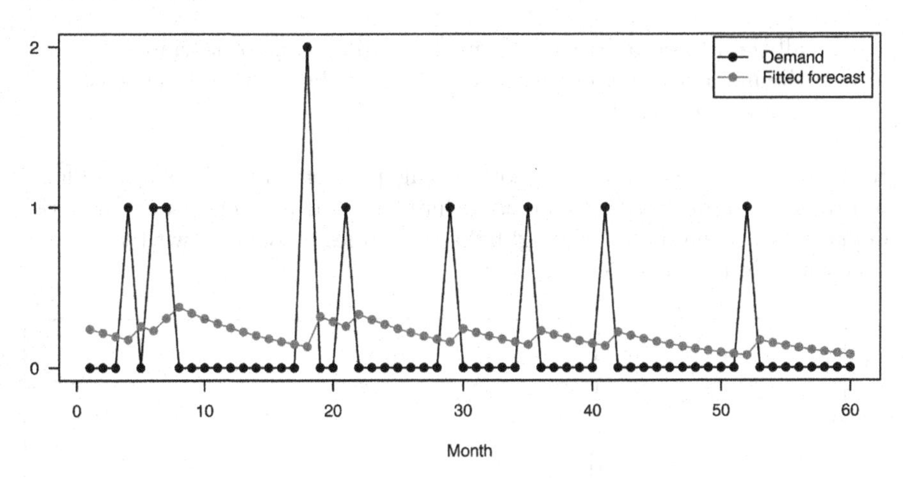

FIGURE 12.4
Single Exponential Smoothing applied to an intermittent time series

This form of forecasting does not make sense in two critical situations. First, consider a context where our intermittent demand is driven by a few customers who replenish this SKU when needed. In this case, we would need a higher forecast (and not a lower forecast) after a long string of zero demands because it becomes *more* likely that these customers will place an order again as more time passes. Second, suppose we have a context where the intermittent demand is driven by *many* customers buying independently (but rarely). In that case, the forecast should not exhibit any time dynamics because, at any point in time, an equally large pool of customers may soon demand the product, even after one particular customer has just bought it.

Another problem with applying Exponential Smoothing to intermittent demands is that we typically make replenishment or production decisions right after a sale depletes our stock. So forecasts that are biased high right after sales will lead to particularly high reorder quantities and unnecessarily high inventories. To overcome these problems, we will now focus on a method to overcome these challenges.

12.3 Croston's method

Croston (1972) examines the problem of forecasting intermittent demand and proposes a specific solution to this problem. This solution is by now an industry standard and bears Croston's name. Instead of Exponentially Smoothing the raw demands, we separately smooth two different time series:

1. all the non-zero demands from the original time series and
2. the number of periods with zero demands between each instance of non-zero demand.

In a sense, we decompose the forecasting problem into the sub-problems of predicting how frequently demand happens and predicting how high the demand is if it happens. We highlight the values associated with these two time series in Figure 12.5.

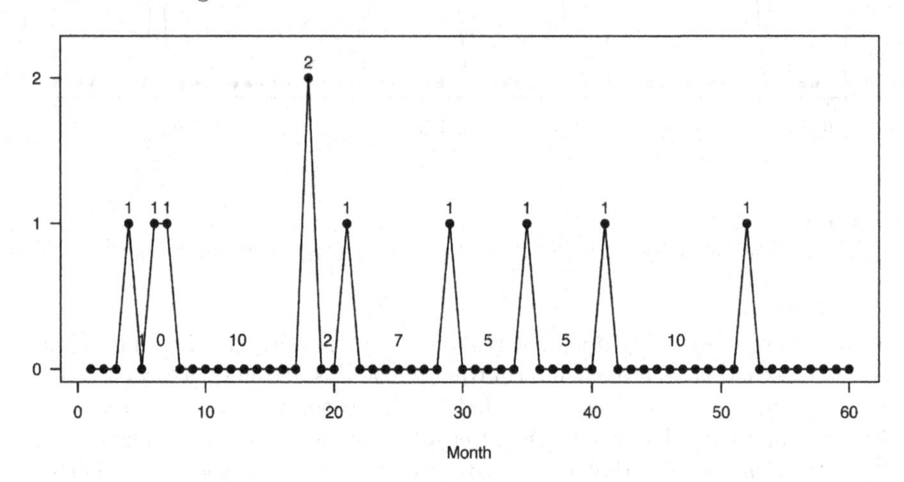

FIGURE 12.5
Demand and time periods without demand in an intermittent series

Thus, the problem becomes one of *separately* forecasting the series of non-zero demands, that is, (1, 1, 1, 2, 1, 1, 1, 1, 1), and the series of periods that are zero in-between demands, that is, (1, 0, 10, 2, 7, 5, 5, 10) in Figure 12.5. Croston's method applies Exponential Smoothing to both these series separately (usually with the same smoothing parameters). Further, we only update the two Exponential Smoothing models and forecasts for these two series whenever we observe a non-zero demand.

Let us assume that smoothing the non-zero demands yields a forecast of q, while smoothing the numbers of zero demand periods yields a r. This means that we forecast non-zero demands to be q, while we expect such a non-zero demand once every r periods on average. Then, the demand point forecast in each period is simply the ratio between these two forecasts, that is,

$$Forecast = q/r \tag{12.1}$$

Croston's method works through averages; while it is not designed to predict when a particular demand spike occurs, it essentially distributes the volume of the predicted subsequent demand spikes over the expected periods until that demand spike occurs. One could also think of Croston's method as predicting the ordering behavior of a single client with fixed ordering costs. According to the classic Economic Order Quantity model, a downstream supply chain partner will lump continuous demand into order batches to balance the fixed cost of order/shipping with the variable cost of holding inventory. According to that logic, we could view Croston's method as a way of predicting the demand this downstream supply chain partner sees – that is, to remove the order variability amplification caused by batch ordering from the time series. Figure 12.6 provides an example of how Croston's method fits and forecasts an intermittent demand series.

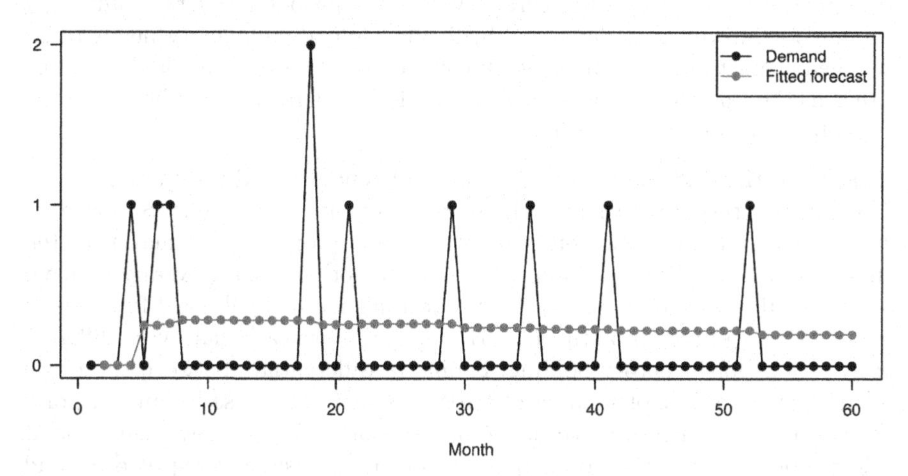

FIGURE 12.6
Croston's method applied to an intermittent demand time series

Croston's method provides some temporal stability of forecasts. Still, it does not address the problem that, for a small number of customers, long periods of non-ordering should indicate a higher likelihood of them placing an order. Further, Croston's method works by averaging and forecasting the *rate* at which demand comes in, not by predicting when demand spikes occur; this makes it hard to interpret forecasts resulting from Croston's method for decision-making. The forecast may say that *on average* over the next five weeks, we will sell one unit each week – where there is likely only one week during which we sell five units.

A closer theoretical inspection shows that Croston's method suffers from a statistical bias (Syntetos and Boylan 2001). Technically, this bias occurs because Equation (12.1) involves taking expectations of random variables, and because the expectation of a ratio is not equal to the ratio of the separate expectations. Modern forecasting software has proposed and implemented various correction factors to compensate for this problem. One example of such a correction procedure is the Syntetos–Boylan approximation (Syntetos and

Boylan 2005; Teunter and Sani 2009), which has often led to better inventory positions (Syntetos, Babai, and Gardner 2015).

Croston's method is straightforward and may appear "too simple." Why is it nevertheless very often used in practice? One factor is its simplicity (compare Chapter 8) – we can explain it quickly, and the logic underlying the method is intuitive to understand. Another explanation may be that intermittent demand often does not exhibit many dynamics that can be modeled. It is tough to detect seasonality, trends, or similar effects in intermittent demands, so trying to create a more complex model quickly runs into overfitting issues (see Section 11.7). The literature nevertheless proposes several competing models for intermittent demands. Still, their added complexity needs to be weighed against any gain in accuracy compared to Croston's method – and no other method has so far consistently outperformed Croston's method with the Syntetos–Boylan approximation.

That said, there is one situation in intermittent demand forecasting where Croston's method performs poorly, namely, for lumpy demands. Suppose we have a demand of one unit once every 10 weeks (non-lumpy demand) or ten units once every 100 weeks (lumpy demand). In both cases, the average demand is $1/10 = 10/100 = 0.1$ unit per week. Croston's method will yield forecasts of about this magnitude in both cases. However, the two cases have very different implications for inventory holding. Failing to differentiate between these two situations is not a shortcoming of Croston's method per se because a point forecast of 0.1 is an entirely accurate summary of the long-run average demand. The problem is that the point forecast does not consider the spread around the average.

Unfortunately, there is no commonly accepted method for forecasting lumpy demands for inventory control purposes. Most forecasters use an ad hoc method like "stock up to the highest historical demand" or a standard forecasting method with high safety stocks. They may also try to reduce their reliance on forecasting for such time series by managing demand and moving to a make-to-order system as much as possible. The challenges of intermittent demand forecasting are one reason why spare parts inventory systems (generally with intermittent demand patterns) increasingly utilize additive manufacturing to print spare parts on demand (D'Aveni 2015).

Standard forecast accuracy metrics can exhibit very surprising behavior when applied to intermittent or lumpy time series. For instance, the MAE will very often be smallest for a flat zero forecast. Thus, paying attention to error metrics is even more important in intermittent or lumpy contexts. See Section 17.4 for details.

Croston's method is implemented in the `forecast` (Hyndman et al. 2023) and `fable` (O'Hara-Wild et al. 2020) packages for R and in the `croston` package for Python (Mohammadi 2022).

Key takeaways

1. Count data and intermittent demands are probably responsible for only 20% of your sales but may account for 80% of your inventory costs. Therefore, it makes sense to invest in forecasting them well.
2. Do not use Exponential Smoothing or ARIMA to forecast intermittent demands. Instead, use Croston's or other methods dedicated to intermittent series. Pay particular attention to the connections between forecasts and inventory control.
3. If you can expand the time unit or aggregate across locations, you can often convert intermittent time series into non-intermittent series that are easier to forecast.
4. Lumpy demands are challenging to forecast since the average rate may be useless for inventory control.
5. Forecast accuracy metrics can be very misleading in intermittent or lumpy contexts. Be careful!

13

Forecasting hierarchies

Forecasts inform many organizational decisions, on inventories, capacity, staffing, cash flow, and budgets. These decisions happen at different organizational levels (e.g., firm vs. division vs. unit) and in different time granularities (e.g., daily, monthly, quarterly, or yearly). Accordingly, time series have natural *hierarchical* structures in "structural" and temporal dimensions (Seaman and Bowman 2022; Syntetos et al. 2016). If we simply forecast each hierarchy level separately, then the forecasts will typically not be *coherent*: the sum of lower level forecasts will not be equal to the higher level forecast, although *historical* sales are of course coherent. In addition, lower-level time series are detailed but noisy, whereas higher level series are less noisy but can lack detail. The present chapter addresses forecasting in hierarchies and discusses various ways of achieving coherence (and also discusses when coherence is *not* called for).

13.1 Structural hierarchies

The first kind of hierarchy that comes to mind is the *structural* kind, which comes in three main variants: (a) organizational, (b) location or geographical and (c) product hierarchies. In the organizational dimension, a company is typically divided into different divisions that address different market segments. In the location hierarchy, we may have groupings of activities by country or region, or our supply chain may be regionally organized. Figure 13.1 gives a simple example of a location hierarchy. Finally, in the product dimension, products are usually organized in a product hierarchy that groups Stock Keeping Units (SKUs) into products or product groups, these into brands, and these again possibly into categories – where the precise categorization differs very much by industry.

Bills of materials also create hierarchical structures among our products: Suppose we bake both blueberry and chocolate chip muffins. We need forecasts for the demands for both muffin types separately to plan production, but when sourcing flour, we only need a forecast of our total flour demand. We do not care which kind of muffin we use a particular pound of flour for. The historical consumption of flour equals the sum of the consumptions for the two different

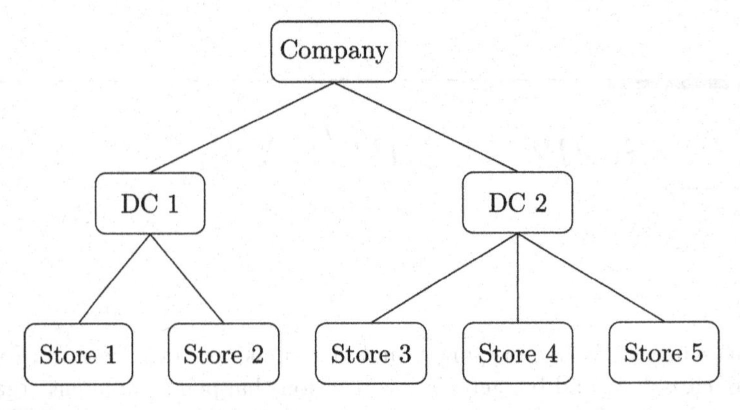

FIGURE 13.1
A simple location hierarchy for a company with two distribution centers (DCs) serving five stores

muffin types. But if we separately forecast flour demand for each muffin type and total flour demand, the two lower level forecasts will not add up to the higher level one.

Product proliferation and the creation of more and more product variants tends to increase the complexity of the product hierarchy, and market differentiation may have similar effects on organizational or location hierarchies.

Structural hierarchies may be *nested* or *crossed*. An example of a nested hierarchy would be if the product hierarchy is nested within the organizational hierarchy of the company – or in other words, the company is organized by product considerations. However, if all products are sold in different geographies, then the location and the product hierarchy are crossed. In a business-to-business context, we also often deal with specific customers or accounts, which will likely go across products, possibly even across our geographical and organizational hierarchy.

We come across challenges when the manufacturing division requires a forecast per SKU across all geographies. While an account manager needs a forecast of demand from their main accounts within a geography but across all products, a marketing manager asks for a forecast of total demand for a single brand within a geography, across all products in that brand. A brand manager at a Consumer Packaged Goods company plans marketing campaigns that affect all SKUs in a brand; a distribution center (DC) needs to plan capacity and staffing to support the stocking and movement of all SKUs assigned to the DC. Demand planners work with product-level demand data, but financial planners report revenues at the division or company level. A retail company may need store-specific forecasts for particular SKUs to determine the store replenishment policy, but it will also need forecasts for the same SKUs at its DC, which serves many stores, to plan the replenishment of the DC. In business-to-business settings, salespeople may need customer-specific forecasts, whereas production planners need a forecast of the total demand across all customers. A consumer goods manufacturer may be interested in sales of a specific SKU to retailer A in region X and to retailer B in region Y, crossing the geographical and customer hierarchies.

Of course, we could always pull the historical data appropriate for each forecast and fit a model to this data. However, if we calculate all such forecasts separately, they will be almost completely independent and will not be guaranteed to be coherent, and we will not be leveraging "similar" slices through the data cube in any way. We will consider different methods of addressing these issues in the next section.

13.2 Forecasting hierarchical time series

Suppose we ignore the hierarchical and interdependent nature of our forecasting portfolio. In that case, we could take the time series at the different aggregation levels – say, sales of SKU A, B and C, and total sales in the category containing A to C – as independent time series and forecast them separately. The problem with this approach is that the forecasts will not be coherent. Whereas the historical demands of the different SKUs add up to the total historical category demand, the forecasts of the different SKUs will almost certainly not add up to the forecast on the category level. "The sum of the forecasts is not equal to the forecast of the sum."

There are various ways of addressing this problem, all with advantages and disadvantages, which we will describe next. Figure 13.2 shows a simple example product hierarchy consisting of three SKUs, each with three years of simulated history and one holdout year. In the following sections, we will work with this hierarchy and show coherent forecasts derived in different ways.

Bottom-up forecasting

The simplest form of hierarchical forecasting is the *bottom-up* approach. This method works just as it sounds: we take the time series at the lowest hierarchical level, forecast each of these series separately, and then aggregate the forecasts up to the desired level of aggregation (without separately forecasting at the higher levels). Many firms use this method without explicitly identifying it as such. The total demand forecast is the sum of product forecasts – a simple bottom-up procedure.

Bottom-up forecasting also works in situations where we do not simply sum up lower level time series but require multiplers, as in bills of material. In the

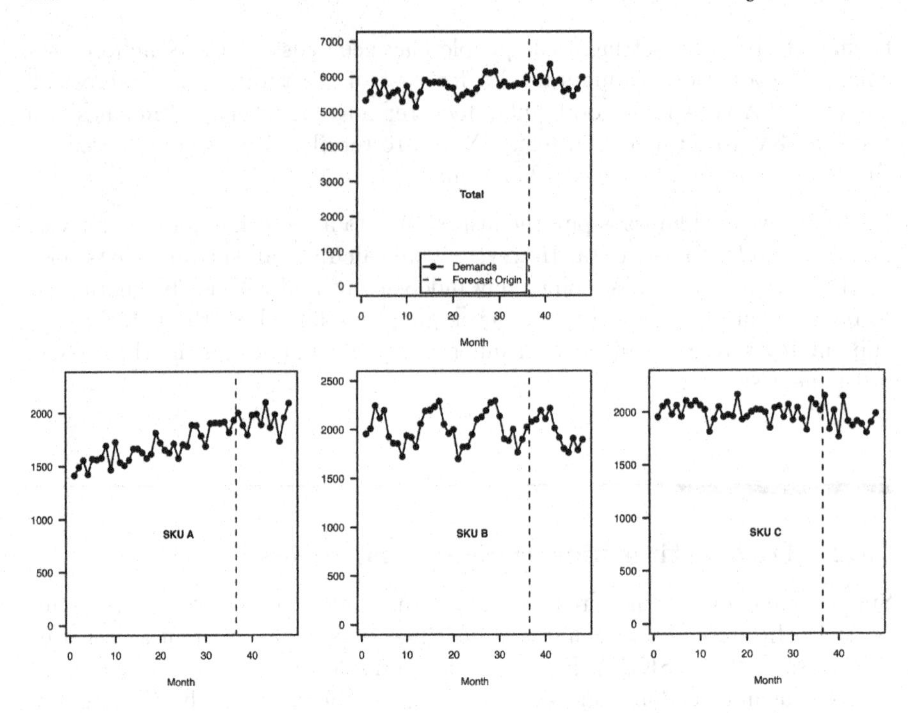

FIGURE 13.2
An example hierarchy with three SKUs' time series

muffin-baking example above, the bakery would forecast its sales of blueberry and chocolate chip muffins separately. It then multiplies each forecasted number with the specific quantity of flour going into each separate muffin and sums the separate forecasts for flour to get a final aggregate flour demand forecast. This is an example of bottom-up forecasting with conversion factors (flour volume per dozen muffins).

One advantage of this approach lies in its simplicity. Implementing this method is effortless, and the bottom-up forecasting process offers little opportunity for mistakes. In addition, it is also straightforward to explain this approach to a non-technical audience. Further, the method is very robust: even if we badly misforecast one series, we limit our error to this series and the levels of aggregation above it – the grand total will not be perturbed a lot. In addition, bottom-up forecasting works very well with causal forecasts since you are more likely to know the value of your causal effect, like the price, on the most fine-grained level. Finally, bottom-up forecasting can lead to fewer errors in judgment for non-substitutable products, increasing the accuracy of forecasts if human judgment plays an essential role in the forecasting process (Kremer, Siemsen, and Thomas 2016).

However, bottom-up forecasting also offers challenges. We lose reliability as we "zoom in" and may miss the forest among the trees. The more granular and disaggregated the historical time series are, the more intermittent they will be, especially if we cross different hierarchies. On a single SKU × store × day level, many demand histories will consist of little else but zeros. On a higher aggregation level, for example, SKU × week, the time series may be more regular. Forecasts for intermittent demands are notoriously tricky, unreliable, and noisy (see Chapter 12). Second, while some causal factors are well defined on a fine-grained level, general dynamics are tough to detect. For instance, a set of time series may be seasonal, but the seasonal signal may be weak and difficult to detect and exploit. Seasonality may be visible on an aggregate level but not at the granular level. In such a case, fitting seasonality on the lower level may even decrease accuracy (Kolassa 2016b).

Figure 13.3 shows bottom-up forecasting applied to the example hierarchy from Figure 13.2. The trend in SKU A and the seasonality in SKU B are captured well, and the forecast on the Total level consequently shows *both* a weak trend and weak seasonality. However, the total forecast is too high, since the trend forecast for SKU A overshot its future demand.

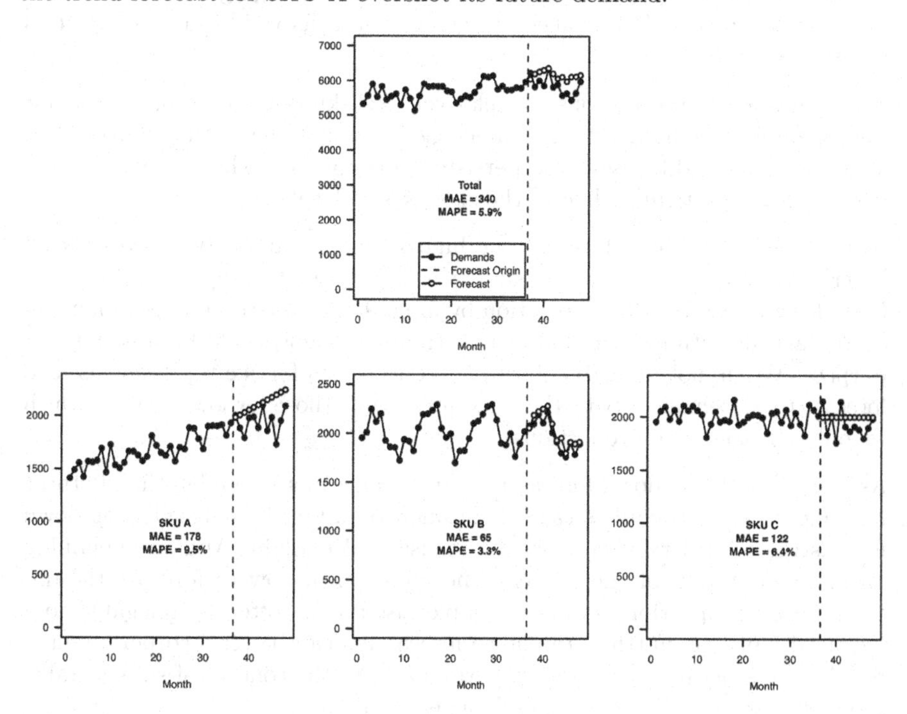

FIGURE 13.3
Bottom-up forecasts

Top-down forecasting

Top-down is a slightly more complex approach than bottom-up forecasting. As the name implies, this process means that we forecast at the highest level of the hierarchy (or hierarchies), then disaggregate forecasts down to lower levels.

What is more complex about top-down forecasting? We need to decide how to disaggregate the higher-level forecasts. And this, in turn, is nothing more than forecasting the proportions in which, say, different SKUs will make up total category sales in a given month. Of course, these proportions could change over time.

One way of forecasting proportions is to disregard possible changes in proportions over time. Following this approach, we could disaggregate the sum by calculating the proportions of all historical sales. This approach is also called "disaggregation by historical proportions" and assumes that historical proportions will continue to be valid in the future. We could instead use only the most recent proportions to disaggregate our forecasts ("disaggregation by recent proportions"). Doing so would allow us to be more responsive to changes in our data. Such an approach assumes constantly changing proportions. The debate from Section 7.3 on stability vs. change is as valid in this context as before.

Sometimes, these proportions are relatively well-known and stable quantities. For example, the distribution of shoe sizes does not change much over time. Thus, we can readily use these persistent proportions when disaggregating demand for a particular shoe to the shoe × size level.

Alternatively, we could forecast the higher level as above but also forecast lower-level demand series and break down total forecasts proportionally to the lower-level forecasts ("disaggregation by forecasted proportions"). Any number of forecast algorithms, seasonal or not, trended or not, could be used for this purpose. We do not need to determine the forecasts for the top level (used to forecast total demands) with the same method as those for lower levels (which we only use to derive proportions).

As you can tell, top-down forecasting can become more complex than bottom-up forecasting and requires careful consideration. On the plus side, top-down forecasting is still relatively simple to use and explain. And in explaining an implementation of the process, the question of how to forecast the disaggregation proportions (a more complex issue) can often be relegated to a technical footnote. Further, top-down forecasting can better incorporate interrelationships between time series, particularly in the context of substitutable products (Kremer, Siemsen, and Thomas 2016).

On the downside, the other advantages and disadvantages of top-down forecasting mirror those of the bottom-up approach. Whereas bottom-up forecasting is robust to single misforecasts, top-down forecasting faithfully pushes every error on the top level down to every other level.

Furthermore, top-down forecasting often gets the total forecast right at the expense of lower-level errors. How this process breaks the sum into its parts depends on how we forecast future proportions. Suppose demand gradually shifts from SKU A to SKU B because of changes in customer taste. In that case, we need to include this shift explicitly in our proportion estimates, or the disaggregated forecasts will not benefit from this critical information.

While causal factors like prices are well defined for bottom-up forecasting (although their effect may be hard to detect, see Kolassa 2016b), they are not necessarily as well defined for top-down forecasting. Prices are defined on single SKUs, not on "all products" level. We could use averages of causal factors, for example, average prices in product hierarchies. However, if we use unweighted averages, we relatively overweight slow-selling products. Suppose we use weighted averages of prices, for example, weighting each SKU's price with its sales to account for its importance in the product hierarchy. In that case, we have to solve the problem of how to calculate these weights for the forecast period. Calculating a weighted average price may be tempting, where we weigh each SKU's price with its forecasted sales. However, such an approach would be putting the cart before the horse, since forecasting the bottom levels is exactly what we are trying to do.

Figure 13.4 shows top-down forecasting, using disaggregation by historical proportions, applied to the example hierarchy from Figure 13.2. As the trend and seasonal signal are weak at the Total level, the automatic model selection picks Single Exponential Smoothing. Consequently, neither the total forecast nor the disaggregated forecasts on the SKU level exhibit trend or seasonality. Top-down forecasting is inappropriate in this example because the bottom-level time series show obvious but different signals (trend in SKU A, seasonality in SKU B, nothing in SKU C). If the bottom-level series exhibit similar but weak signals, top-down forecasting may perform better than bottom-up (Kolassa 2016a).

Middle-out forecasting

Middle-out forecasting is a middle road between bottom-up and top-down forecasting. Pick a "middle" level in your hierarchy. Aggregate historical data up to this level. Forecast. Disaggregate the forecasts back down, and aggregate them up as required.

This approach combines both the advantages and disadvantages of bottom-up and top-down forecasting. Demands aggregated to a middle level in the hierarchy will be less sparse than on the bottom level, leading, e.g., to better-defined seasonal signals. It may also be easier to derive aggregate causal factors when going to a middle level than when going to the top level. For

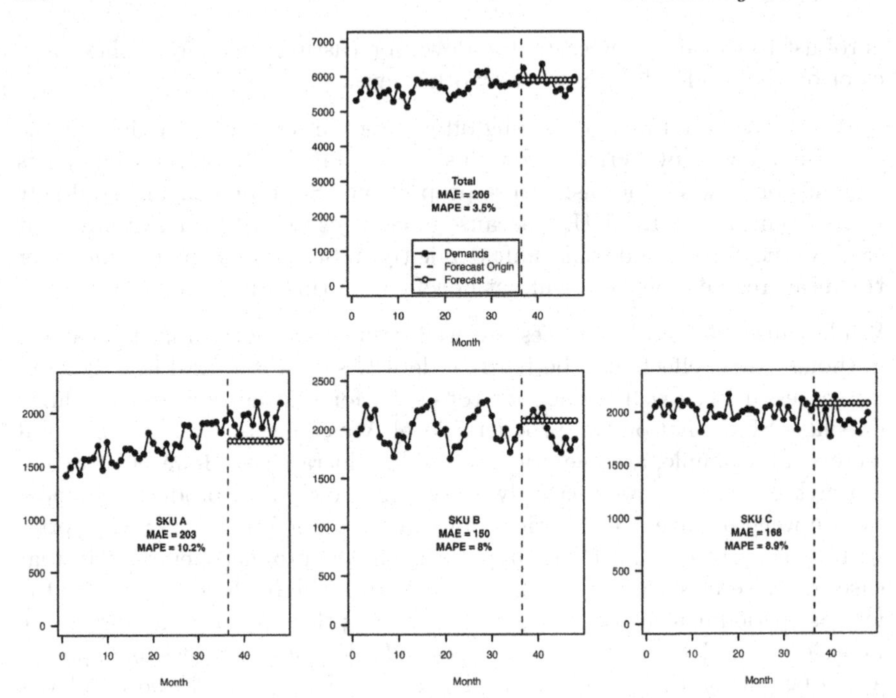

FIGURE 13.4
Top-down forecasts

instance, if all products of a given brand have a "20% off" promotion, we can aggregate historical demands to total brand sales and apply a common "20% off promotion" predictor to the aggregate. (However, as above, we would need to explicitly model *differential* sensitivity to price reductions for the different SKUs in the brand – a simple approach would yield the same forecasted uplift for all products.)

Once we have the forecasts on a middle level, aggregating them up is just as simple as aggregating bottom-level forecasts in a bottom-up approach. And conversely, disaggregating middle-level forecasts down to a more fine-grained level is similar to disaggregating top-level forecasts in a top-down approach: we again need to decide on how to set disaggregation proportions, for instance, (the most straightforward way) doing disaggregation by historical proportions.

Optimal reconciliation forecasting

One recently developed approach, and an entirely novel way of looking at hierarchical forecasting, is the *optimal reconciliation* approach (Hyndman et al. 2011; Hyndman and Athanasopoulos 2014). The critical insight underlying this approach is to return to the consistency problem of hierarchical forecasting.

If we separately forecast all our time series on all aggregation levels, then the point forecasts will not be coherent. They will not "fit" together. In that case, we can calculate a "best (possible) fit" between forecasts made directly for a particular hierarchical level and indirectly by either summing up or disaggregating from different levels, using methods similar to regression. We will not go into the statistical details of this method and refer interested readers to Hyndman et al. (2011).

Calculating optimally reconciled hierarchical forecasts provides several advantages. One benefit is that the final forecasts use all component forecasts on all aggregation levels. We can use different forecasting methods on different levels or even mix statistical and judgmental forecasts. Thus, we can model dynamics on the levels where they are best fitted, for example, price influences on a brand level and seasonality on a category level. We, therefore, do not need to worry about making hard decisions about how to aggregate causal factors. If it is unclear how to aggregate a factor, we can leave it out in forecasting on this particular aggregation level. And it turns out that the optimal reconciliation approach frequently yields better forecasts on all aggregation levels, beating bottom-up, top-down, and middle-out in accuracy because it combines so many different sources of information.

Optimal reconciliation, of course, also has drawbacks. One is that the specifics are harder to understand and communicate than the three "classical" approaches. Another disadvantage is that while it works very well with "small" hierarchies, it quickly poses computational and numerical challenges for realistic hierarchies in demand forecasting, which could contain thousands of nodes arranged on multiple levels in multiple crossed hierarchies. For single (non-crossed) hierarchies or for crossed hierarchies that are all of height 2 ("grouped" hierarchies), one can do clever algorithmic tricks (Hyndman, Lee, and Wang 2016). However, the general case of crossed larger hierarchies is still intractable, and optimal reconciliation was not used by many teams in the recent M5 forecasting competition (Makridakis, Spiliotis, and Assimakopoulos 2022), presumably because of exactly these computational difficulties.

Figure 13.5 shows optimal combination forecasting applied to the example hierarchy from Figure 13.2. As in bottom-up forecasting (Figure 13.3), the method captures the trend in SKU A and the seasonality in SKU B well. The forecast on the aggregate level consequently shows both a weak trend and weak seasonality. Note that errors for optimal combination forecasts are even lower than those for bottom-up forecasts. This is a frequent finding. Optimal reconciliation tends to be more accurate.

Optimal reconciliation forecasting for hierarchical time series is implemented in the `hts` package for R (Hyndman et al. 2021).

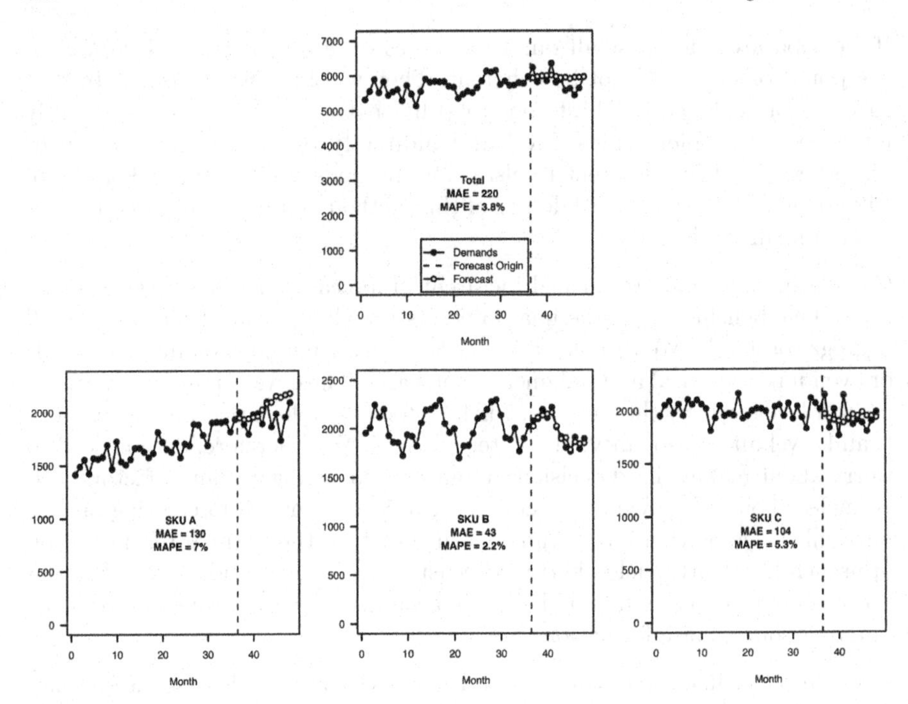

FIGURE 13.5
Optimal reconciliation forecasts

Other aspects of hierarchical forecasting

A hierarchy used for forecasting should follow the many-to-one rule: many lower levels group into one higher level, but a lower level should not be part of several higher levels. Forecasting hierarchies are sometimes inherited from other organizational planning processes that do not follow this many-to-one rule. As a result, a product can be, for example, sometimes a part of multiple categories. This issue can lead to double counting and inconsistencies when we sum lower levels into higher levels or disaggregate higher levels into lower levels.

If it looks like one hierarchical node might be a child of multiple parent nodes, that may be a symptom of more than one hierarchy being involved: if a car manufacturer customer active in Germany and the UK is member of both the "automotive" and the "Europe" nodes in our single organizational hierarchy, then the best approach is to explicitly create separate (but crossed) customer and geographical hierarchies.

Hierarchical structures allow forecast improvements even if we are not interested in hierarchical forecasts per se. For example, we may only need SKU-level forecasts but still be interested in whether product groups allow us to improve these SKU-level forecasts. For instance, estimating multiplicative seasonality components at a higher level is often helpful if we are confident that all lower levels should exhibit this seasonality. These time series components are much more apparent at that higher level. Once estimated, we can apply them directly to forecast (deseasonalized) lower-level series. In other words, while seasonality is estimated top-down, the remaining forecast (factoring in these seasonal factors) can be done bottom-up. Mohammadipour, Boylan, and Syntetos (2012) explain this approach in more depth. We could use the same method to calculate the effects of trends, promotions, or any other dynamic on an aggregate level, even if we are not interested in aggregate forecasts per se.

One crucial point to remember when forecasting hierarchical data is that structural hierarchies are often not set up with forecasting in mind. For instance, products grouped by suppliers may sort different varieties of apples into different hierarchies, although they appear the same to customers and would profit from hierarchical forecasting as discussed below. One retailer may divide his fruit and produce first into organic vs. conventional and then into the different varieties of fruit, whereas another retailer may first group by fruit and divide organic vs. conventional bananas only at the bottom of the hierarchy. Since the organic and conventional varieties of the same fruit may share similar seasonal patterns, making the organic vs. conventional distinction lower in the hierarchy may help improve forecasts, because then the two varieties of banana are closer together in the hierarchy and can share information. It may thus be worthwhile to create dedicated "forecasting hierarchies" and to check whether these improve forecasts (Mohammadipour, Boylan, and Syntetos 2012).

13.3 Temporal hierarchies and aggregation

Increasing computing power and improvements in database architectures allow us to capture and store data in increasingly finer temporal granularity, such as point-of-sales transaction timestamps in a supermarket and patients' exact arrival times in a hospital. While point-of-sales systems record sales the second they are made, we require forecasts at coarser time granularities. Generally, the time unit registered will not match the time unit necessary for decision-making and planning. We thus often first have to translate the raw time series of recorded data at higher frequencies (e.g., daily/hourly) into lower frequency forecasts (e.g., weekly/monthly totals).

We then face the challenge that different decisions in the company require forecasts on different temporal granularities. For example, a supply chain manager at a regional warehouse may make weekly decisions for distribution, monthly decisions for stock replenishment, and yearly decisions for purchasing. At a brick and mortar retailer, the store manager needs hourly forecasts for planning production at the deli counter and daily forecasts for replenishing the shelves, whereas the promotion planner looks at weekly total forecasts when planning offers. An online fashion retailer may need forecasts of daily goods movements to schedule the warehouse workforce and will also need forecasts for the entire fashion season, ranging from a few weeks to a few months, to procure products in the countries of origin.

Thus, temporal hierarchies are very similar to the structural hierarchies discussed above, and forecasting in temporal hierarchies can be addressed in similar ways – almost. There are two key differences. One is that while structural hierarchies are only present if we have an entire *collection* of time series, along with their structural hierarchy, a temporal hierarchy is already implicitly given in a single time series: we can consider forecasts on various time granularities even if we have only one series to forecast. The other main difference between temporal and structural hierarchies is the one between *non-overlapping* and *overlapping* temporal aggregation: we can aggregate data in consecutive non-overlapping buckets, e.g., monthly data to quarterly totals (see Figure 13.6), but we can also aggregate it into overlapping totals, say of 3-month periods (see Figure 13.7). It turns out that modeling overlapping temporal aggregates may even improve forecasts if we are not interested in the overlapping buckets as an end result (Boylan and Babai 2016; Rostami-Tabar, Babai, and Syntetos 2022).

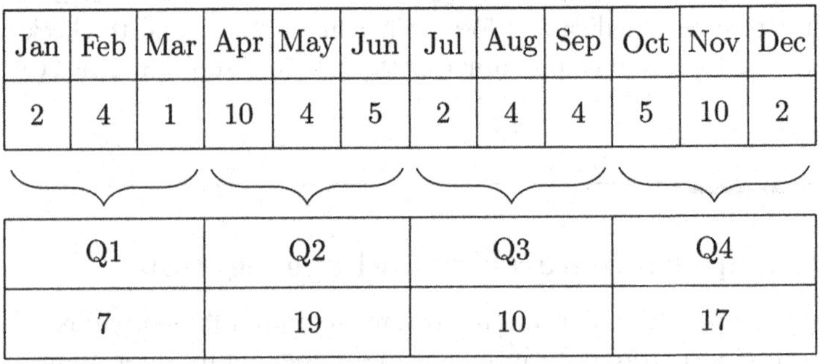

Jan	Feb	Mar	Apr	May	Jun	Jul	Aug	Sep	Oct	Nov	Dec
2	4	1	10	4	5	2	4	4	5	10	2

Q1	Q2	Q3	Q4
7	19	10	17

FIGURE 13.6
Non-overlapping temporal aggregation of monthly demands to quarters

Figure 13.8 illustrates a time series of monthly granularity and non-overlapping aggregated series at quarterly and annual levels for retail sales in millions of Australian dollars (Godahewa et al. 2021). Seasonality and trend dominate the monthly and quarterly series. As we increase the granularity to the annual level, the seasonality disappears, while the trend is very apparent.

Applying such a non-overlapping temporal aggregation approach can influence the components of a time series, including seasonality, trend, and autocorrelation. In other words, different aggregation levels can reveal or conceal the various time series components. We generally observe fewer peaks and troughs

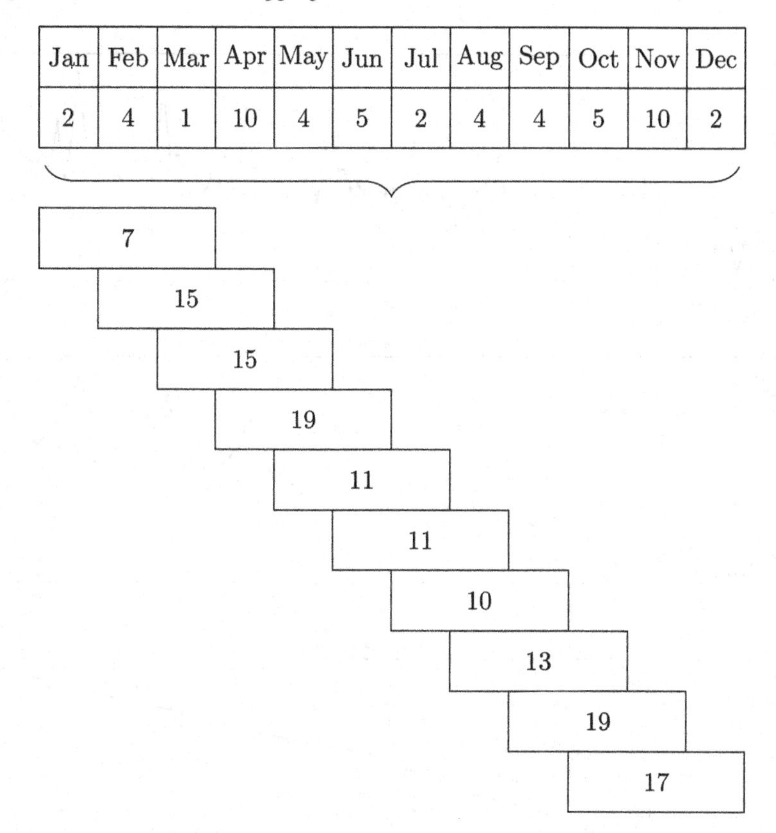

FIGURE 13.7
Overlapping temporal aggregation of monthly demands to 3-month buckets

in low-frequency (aggregated) series than in high-frequency series, i.e., a lower frequency reduces the impact of seasonality. Therefore, while seasonality may be a dominant feature in a higher frequency (e.g., daily) time series, this seasonal pattern can disappear by increasing the aggregation level to a lower frequency (e.g., yearly). The strength of a trend may increase by aggregating the series. We are more likely to see trend patterns in lower-frequency time series. Moreover, autocorrelation patterns can decrease with aggregating series to a lower frequency.

In the case of intermittent demand time series (see Chapter 12), moving from a higher to a lower frequency reduces or even entirely removes the time series intermittency (Nikolopoulos et al. 2011), minimizing the number of periods with zero demands. This form of aggregation can allow the use of conventional forecasting methods in otherwise intermittent series. Figure 13.9 illustrates a time series of spare part sales (Godahewa et al. 2021) at the original monthly and aggregated quarterly and annual levels. We can see how non-overlapping

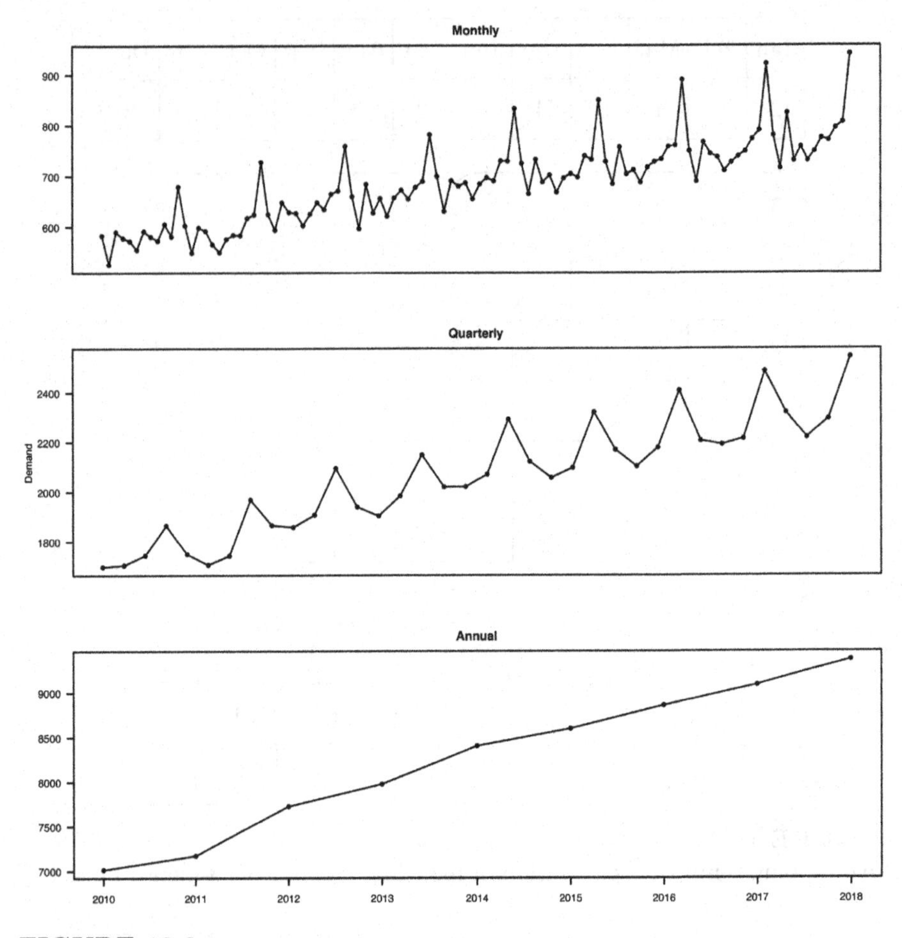

FIGURE 13.8
Australia retail sales in millions of Australian dollars. Original time series sampled at a monthly frequency and aggregated at quarterly and annual levels

temporal aggregation transforms the time series, leading to fewer zero values for the quarterly and almost no zero values for the annual time series.

Forecasts are always necessary at the time unit they are needed for decision-making. But, for example, if we need weekly forecasts and daily data is highly intermittent, a possible approach is to aggregate the daily data into weekly data before forecasting and finally disaggregating the forecast again. Non-overlapping temporal aggregation can thus have its advantages.

We emphasize a few limitations of using a single time unit to generate forecasts:

- The forecast generated at the time unit necessary for decision-making is not necessarily the most accurate one.

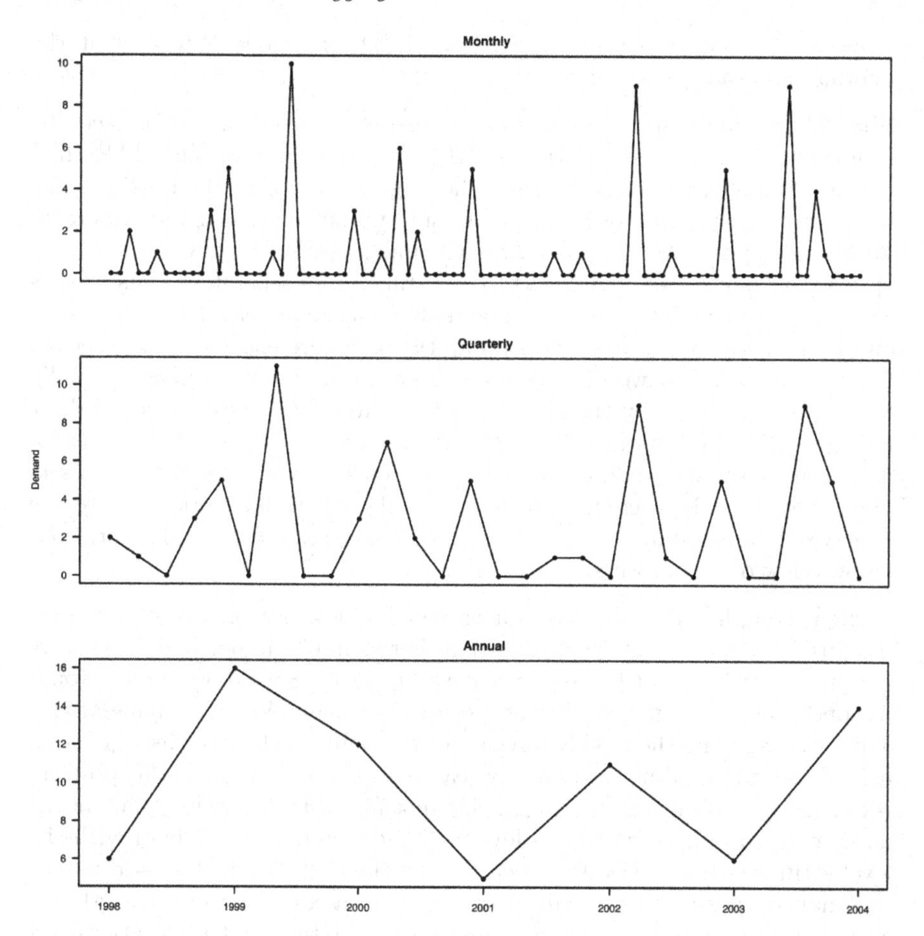

FIGURE 13.9
A monthly intermittent time series at different levels of non-overlapping temporal aggregation

- A single time unit ignores time series components at all other levels and does not benefit from such information.
- Two sets of forecasts generated from the same data at different temporal aggregation levels are rarely consistent. For instance, the sum of daily forecasts is likely different from a weekly forecast based on aggregated weekly data. If these two different forecasts inform different decision-making processes, the "weekly" decisions will not be coherent with the "daily" decisions.

Instead of using a single time unit to generate the forecast, we can create multiple time series using non-overlapping temporal aggregation (e.g., monthly, quarterly, yearly), train and create forecasts at each level, and combine available forecasts at multiple levels of aggregation. The key idea is to exploit the information available at various time levels.

The *Multi Aggregation Prediction Algorithm (MAPA)*, introduced by Kourentzes, Petropoulos, and Trapero (2014), applies this idea. The MAPA first constructs multiple time series from the original data using non-overlapping temporal aggregation. For example, we sum up our daily demands to create weekly demands. We then forecast each time series separately by fitting an appropriate model (see Chapter 9) to each time series separately. This process would give us daily forecasts (based on daily data) and weekly forecasts (based on weekly data). Next, we would sum up the daily forecasts to create a second weekly forecast, and we disaggregate the original weekly forecast (usually by simply dividing it by the number of workdays) into a second set of daily forecasts. This process would leave us with two weekly forecasts and two sets of daily ones. Last, we would combine the two weekly forecasts and the two sets of daily forecasts by averaging them (or calculating the median) to create one final weekly forecast and one final set of daily forecasts. Figure 13.10 provides an overview of this process.

A slight complication can arise for seasonal series. Seasonality can quickly disappear for some levels of aggregation. For example, in our daily series, we may have seasonality (demand is always higher on Saturday), which would disappear once we aggregate the data to weekly demand. When applying MAPA and disaggregating the weekly forecast to the daily level, these disaggregated daily forecasts would not exhibit any seasonality. The solution to this problem is to aggregate/disaggregate components, not forecasts. We would think of the disaggregated daily forecasts as level estimates, and average them with the level estimates from directly forecasting at the daily level, then adding the seasonal component at the daily level to derive a combined forecast. MAPA allows us to capture time series components at each temporal aggregation level. The algorithm also benefits from forecast model combination. While MAPA works well with Exponential Smoothing models (so we can combine all the level, trend, and seasonal components), one can also extend the idea to employ other forecasting models – or even apply it to structural instead of temporal hierarchies.

A related approach is the *temporal hierarchies* process proposed by Athanasopoulos et al. (2017). The idea is to think about the temporal levels as an explicit hierarchy and treat it with the exact same tools as discussed in Section 13.2. Higher frequency time series are at the bottom of this hierarchy (e.g., daily), and lower frequencies (e.g., monthly, quarterly, semi-annual, annual) are at the higher levels. Figure 13.11 shows an example of a temporal hierarchy for a monthly time series with three temporal levels at the monthly, quarterly, and annual levels. For a given forecasting method, we generate forecasts at all levels of the hierarchy and then combine these forecasts to develop reconciled forecasts for all levels.

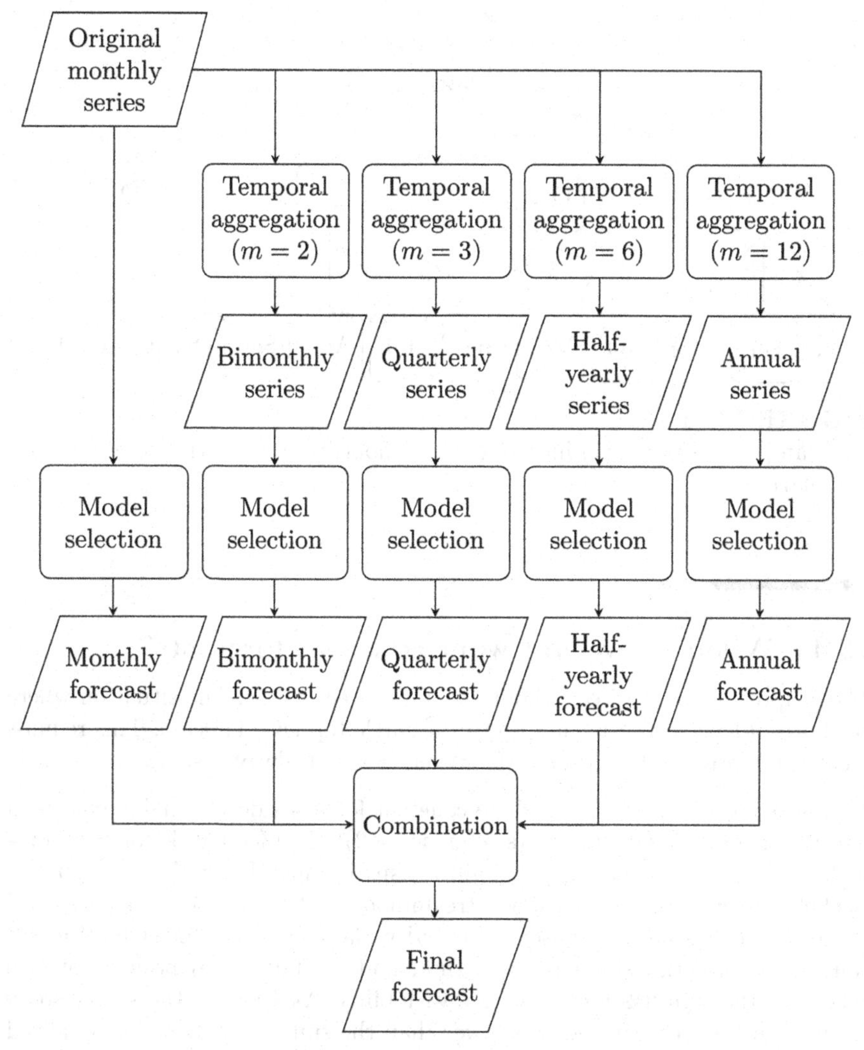

FIGURE 13.10
The Multi Aggregation Prediction Algorithm (MAPA) applied to a monthly series

We will not go into the technical details of these approaches – readers can refer to Athanasopoulos et al. (2017) and Kourentzes, Petropoulos, and Trapero (2014), or look at the `thief` (Hyndman and Kourentzes 2018; Athanasopoulos et al. 2017) and `MAPA` (Kourentzes and Petropoulos 2022, 2016; Kourentzes, Petropoulos, and Trapero 2014) packages for R.

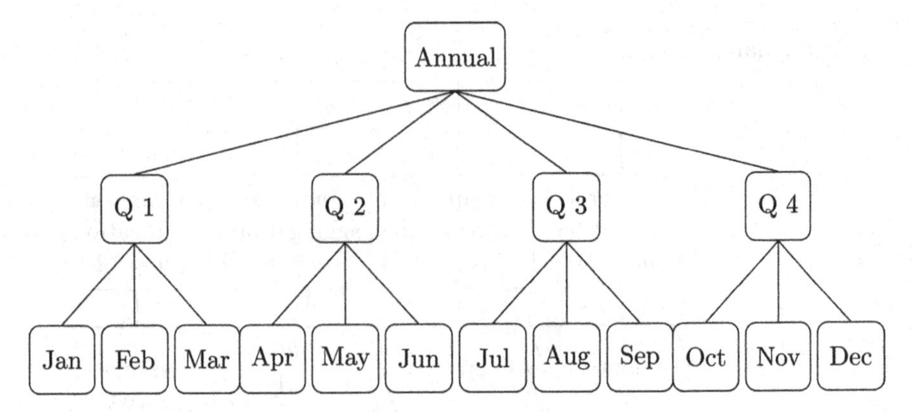

FIGURE 13.11
An example of a temporal hierarchy for a monthly series with three temporal granularities.

13.4 When do we *not* want coherent forecasts?

The chapter so far has been about coherent forecasts, i.e., the situation where the bottom-level forecasts sum up to the single top-level forecast. This requirement is often taken for granted. However, it is not always useful.

The most important example where coherent forecasts do not make sense is in interval or quantile forecasting (see Section 4.2). A 90% quantile forecast for a higher level node in the supply chain is usually much lower than the sum of the 90% quantile forecasts at downstream nodes. Of course, this is as it should be, and it makes perfect sense: noise and variability at the lower level nodes cancel out (sometimes demand is high at node A, but low at node B, or vice versa), so the sum has lower relative variability. As a result, the safety stock required at an upstream node is lower than the sum of safety stocks required at multiple downstream nodes. Thus, it makes no sense to require coherence from prediction intervals or quantile forecasts.

For full predictive densities (see Section 4.3), there is a notion of *probabilistic coherence* (Panagiotelis et al. 2023). However, this is highly technical and an area of active research.

Finally, there are some subtle points about optimizing hierarchical forecasts with regard to certain error measures. Minimizing the Mean Absolute Percentage Error (MAPE; see Section 17.2) is a target that conflicts with the requirement of coherent forecasts, because the MAPE-optimal forecast is not additive (Kolassa 2022c). If your bonus is tied to minimizing MAPEs of coherent forecasts, incentives may not be well aligned.

Key takeaways

1. If you have a small hierarchy and can use dedicated software like R's `hts` package, or have people sufficiently versed in linear algebra to code the reconciliation themselves, go for the optimal reconciliation approach. If the optimal reconciliation approach is not possible, use one of the other (i.e., bottom-up, top-down or middle-out) methods.
2. You can create time series at multiple temporal levels (daily, weekly, monthly), forecast each series, and reconcile these forecasts using optimal reconciliation.
3. If you are most interested in the top-level forecasts, with the other levels "nice to have" or are worried about extensive substitutability among your products, use top-down forecasting.
4. If you are most interested in the bottom-level forecasts, with the other levels "nice to have," or many of your products are not substitutes, use bottom-up forecasting.
5. If you can't decide, use middle-out forecasting. Either pick a middle level that makes sense from a business point of view, for example, the level in the product hierarchy at which you plan marketing activities, or try different levels and look at which one yields the best forecasts overall.
6. Never be afraid of including hierarchical information, even if you are not interested in higher aggregation-level forecasts. It may improve your lower-level forecasts.
7. Consider creating additional "forecasting hierarchies" if that helps to do the job.
8. Do not reconcile hierarchical forecasts blindly. Not all forecasts need to be coherent.

14

Artificial intelligence and machine learning

In this chapter, we will consider "modern" forecasting methods that became more prominent with the advent of Data Science. Umbrella terms used for these methods are "Artificial Intelligence" (AI) and "Machine Learning" (ML). There is no commonly accepted definition of these terms, and boundaries are ambiguous. ARIMA and Exponential Smoothing could certainly be classified as ML algorithms.

A standard categorization in ML distinguishes between *supervised* and *unsupervised learning*. The difference between these two approaches lies in whether training data are labeled or not. While supervised learning examines the relationship among labeled data, unsupervised learning examines patterns among unlabeled data. All forecasting methods represent supervised learning, since we train models to known targets (past observed demands) rather than learning without a known target outcome.

14.1 Neural networks and deep learning

Neural Networks (NNs) are Machine Learning algorithms based on simple models of neural processes. Our brains consist of billions of nerve cells, or neurons, which are highly interconnected. When our brain is presented with a stimulus, e.g., an image or scent of a delicious apple, neurons are "activated." These activated neurons in turn activate other connected neurons through an electrical charge. Over many such stimuli, patterns of connections emerge and change.

NNs (the computer kind, see Figure 14.1) imitate this process. They are initialized with an *architecture* specifying how abstract "neurons" are arranged and interconnected. Typically, neurons are set up in layers, with neurons in neighboring layers connected. An input layer of neurons gets fed input data. This input layer is connected to one or multiple hidden layers. The last hidden layer is connected to an output layer. The number of such layers is called the

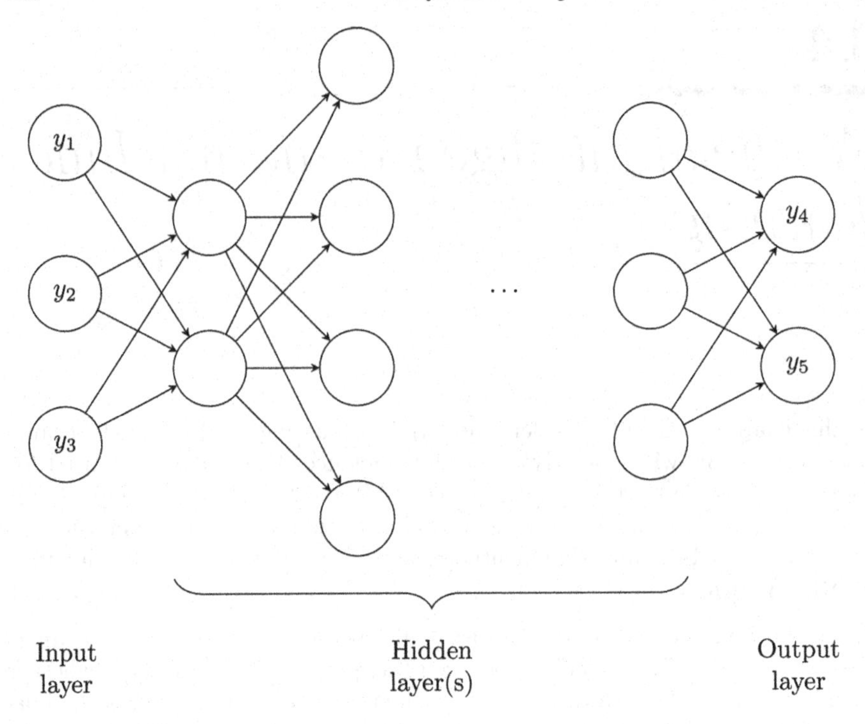

FIGURE 14.1
A neural network

depth of the network. The term *Deep Learning* refers to NNs with many layers, in contrast to simpler architectures with only a few layers.

To train such an NN, we present it with inputs, e.g., the historical observations of a time series, and an output, e.g., the subsequent values of the series. The training algorithm adjusts the strengths of the connections between the neurons to strengthen the association between the inputs and the output. This step is repeated with many input-output pairs in the training data. Once the network has "learned" the association between an input and the corresponding output, we can present it with a new input, e.g., the current value of a time series, and obtain the output the NN associates with that input, e.g., our forecast for a future value of the series. The input to a NN used for forecasting can contain more data than just the historical time series, such as time-constant or time-varying predictor information.

NNs have existed since the 1950s. After some initial enthusiasm, interest waned in these methods as they did not seem to provide practical applications (a period sometimes called an "AI winter"). In recent decades, a combination of theoretical advances in NN training algorithms, more powerful and specialized hardware, and abundant data have fomented an explosion of NN research and applications. NNs are often used in pattern recognition tasks, from image recognition to sound and speech analysis. They can also *generate* patterns, including Deep Fakes of convincing videos, poems, and solutions to academic exam questions. NNs are also increasingly applied to forecasting tasks.

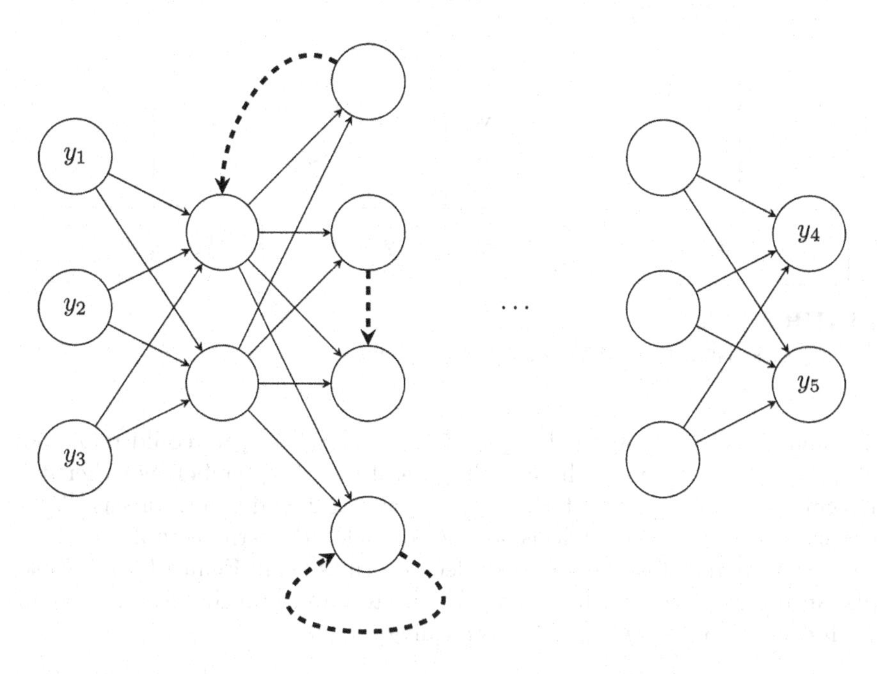

FIGURE 14.2
A recurrent neural network, with recurrent connections highlighted

The art of using NNs for any task lies in setting up the architecture and in pre-processing the data. For instance, it is often helpful in forecasting to connect "later" layers in the NN to "earlier" layers, which results in a *Recurrent Neural Network* (RNN, see Figure 14.2). Such a network can be interpreted as representing "hidden states," i.e., unobserved "states of the world." More sophisticated versions of RNNs can forget information over time. Such an architecture is known as a *Long Short-Term Memory* (LSTM) network. Variants of LSTM are the most common NN architecture used in forecasting.

14.2 Tree-based methods and random forests

Tree-based methods for forecasting (Spiliotis 2022; Januschowski et al. 2022) represent a collection of decision rules. An example of such a decision rule is "*if* the month is July to September, *then* forecast 10, *else* forecast 5". Such

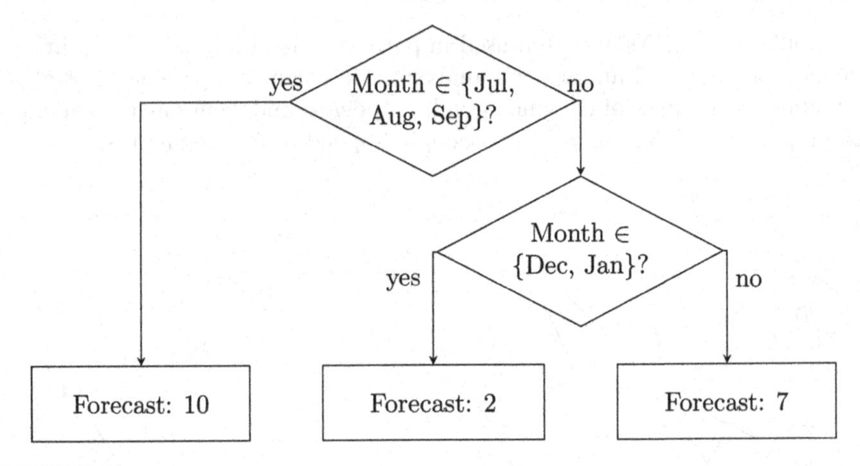

FIGURE 14.3
A simple decision tree to forecast monthly data

decision rules can be applied sequentially. For example, we could refine our decision rule by asking if the month is *not* July to September, whether it is December or January, and if so, we could forecast 2, and if not, forecast 7. We can increasingly refine our decision rules to yield arbitrarily complex decision chains. If we visualize such a set of decision rules, as in Figure 14.3, it looks like an inverted tree, with the root at the top and branching points below. Each decision point yields such a branching.

Such trees – the specific name for these is *Classification And Regression Trees (CARTs)* – are trained, or "grown," by using input and output data. The data are recursively partitioned to minimize the error we would obtain if we used the tree to predict the training data. CARTs overfit easily, i.e., they are prone to follow the noise in the data accidentally. To counteract this, one often limits the maximum depth of the tree or grows it first and then *prunes* it back by cutting off the bottom branches. Alternatively, one can deal with the overfitting issue by using a Random Forest, see below.

Decision trees can process any predictor. In the example above, we used only the month as a predictor. As a result, we modeled monthly seasonality. We could also include the day of the week for daily data, lagged values of the time series itself, or any other kind of predictor – all we need is that the predictor value is known and available for forecasting.

Random Forests (RFs) are a generalization of CARTs. As the name implies, a RF contains hundreds of trees. In addition, some random mutation is involved in growing a RF by picking a random set out of our training data and growing each tree only on a small random subset of the predictors. This strange double randomization is surprisingly effective at creating a diversity of trees. The forecast from an RF is the average of the forecasts from the component trees.

The idea of combining multiple forecasts (here: from the component trees) is an example of ensemble forecasting (see Section 8.6).

14.3 Boosting and variants

We already saw examples of the powerful concept of combining multiple different forecasts via averaging when we encountered Random Forests and Section 8.6. The technique of *Boosting* represents another way of combining various forecasting methods. Specifically, we first model our time series using any method of choice. We then consider the differences between the original time series and the fitted values, i.e., the model errors. In turn, we predict these errors using another method, typically a very simple one (a so-called *weak learner*). Then, we iterate this process. The end model and forecast is the sum of the iterative boosted components. Boosting describes how we use and stack specific models in each step. The idea of Boosting (Schapire 1990) has been around for decades, and several variants have been considered:

- *Gradient Boosting* was proposed by Friedman (2001); here, we replace the differences between the observations and the current fit (the "residuals") with "pseudo-residuals" involving the derivative of the loss function, which makes the subsequent fitting of weak learners easier.
- *Stochastic Gradient Boosting* (Friedman 2002) works similarly to Gradient Boosting but only relies on a random sub-sample of the training data in each iteration. This injection of randomness serves a similar purpose as in Random Forests.
- *(Stochastic) Tree Boosting* is a specific kind of (Stochastic) Boosting, where the weak learners fit in each step are simple decision trees, which we have encountered above. The most straightforward possible tree consists of a single split, which is inevitably called a *stump*.
- *LightGBM* ("Light Gradient Boosting Machine") and *XGBoost* ("eXtreme Gradient Boosting") are specific implementations of Stochastic Tree Boosting, with particular emphasis on efficiency and scalability through distributed learning. These modern methods are frequent top contenders at Kaggle forecasting competitions and dominated the top ranks of the recent M5 time series forecasting competition (Makridakis, Petropoulos, and Spiliotis 2022).

14.4 Point, interval and density AI/ML forecasts

While AI/ML methods are often developed to create point forecasts, these methods can also be used to yield interval and density forecasts. For instance, we can train these models to output quantile forecasts using an appropriate pinball loss function (so-called because it resembles the trajectory of a pinball hitting a wall). Thus, one could train an entire group of NNs, and CARTs, each for a different quantile, and get multiple quantile forecasts and the corresponding prediction intervals. Alternatively, we can use AI/ML methods to forecast the parameters of a probability distribution, which gives us total predictive density. One such setup is DeepAR, which uses a NN to predict the parameters of a discrete probability distribution (Salinas et al. 2020).

14.5 AI and ML versus conventional approaches

A critical difference between AI/ML and conventional forecasting methods, such as ARIMA or Exponential Smoothing, is that the conventional methods are linear. AI/ML models are non-linear. For example, if you take a time series, double each value in the series, model it using conventional methods, and create a forecast, your forecast will also be twice as high as the forecasts you would get from the original series. The conventional methods impose a linear structure on the models they generate. AI/ML methods, in contrast, are more flexible. They can model non-linearities.

Allowing for non-linearities sounds like a valuable feature of AI/ML. However, this flexibility comes at a price. AI/ML methods are prone to overfitting, i.e., they tend to follow the noise instead of detecting the signal in the data. Thus, they are more sensitive to random fluctuations in the data. As a result, AI/ML models can produce forecasts that are also highly variable and, therefore, potentially wildly inaccurate. See Section 11.7 on model complexity, overfitting and the bias-variance tradeoff. To counteract this problem, AI/ML methods are typically fed large amounts of data.

AI/ML methods are very data and resource hungry. They require a lot more data input than traditional forecasting methods. They also need much more processing time to train the model before their forecasts make sense. This can lead to an excessive need for computation hours compared to traditional models, leading to cloud computing costs and energy consumption. According to one estimate, moving a large-scale retail forecasting operation from conventional forecasting methods to AI/ML results in extra CO_2 emissions due to increased energy use in computation that is equivalent to the annual CO_2 production of 89,000 cars (Petropoulos, Grushka-Cockayne, et al. 2022).

If you have a single time series of several years of monthly data, you can easily fit an ARIMA or Exponential Smoothing model to this series and get valuable forecasts. A NN trained on the same series may give highly variable and inaccurate predictions. Thus, AI/ML methods are typically not fitted to a single time series but are instead fitted to a whole set of series. They are *global* methods, in contrast to traditional methods, which are *local* to single time series (Januschowski et al. 2020).

This distinction between global and local methods has two consequences. First, local methods can be easily parallelized for higher performance. Global methods need a lot more work to train efficiently. Second, whereas local methods can be understood in isolation (you can look at and understand an ARIMA forecast from a single series), global methods are tough to understand because the forecast for a given series does not only depend on this series itself but also on all the other series that went into training the NN. Recently, academic and practical work has emphasized the importance of "explainable AI" (XAI). This term emphasizes that AI/ML models should be designed in a way as to help humans understand their outputs and predictions. Relatedly, if we want to "debug" a forecast, it is easier to tweak a local method than a global one. If you modify a global method to work better on a particular series, it may suddenly work worse on a seemingly unrelated other series. Usually, many AI/ML models with similar accuracy exist, and one can use the simplest and easiest to interpret and debug among them – a concept known as *Rashomon Set Theory* (Fisher, Rudin, and Dominici 2019).

In summary, AI/ML has the potential to achieve better forecasting accuracy than more traditional methods through its greater flexibility, but whether this advantage is realized depends on the time series. For instance, in the M5 forecasting competition, AI/ML had a much greater edge on aggregated data than on the more granular SKU $\times$ store $\times$ day level (Makridakis, Petropoulos, and Spiliotis 2022). In addition, while AI/ML methods won the M5 competition, conventional methods were competitive: the simplest benchmark, Exponential Smoothing on the bottom level with trivial bottom-up aggregation to higher hierarchy levels, performed better than 92.5% of submissions (Kolassa 2022a). And the accuracy advantage of AI/ML methods comes at the price of higher complexity, i.e., higher costs. It is not apparent that a more costly, more complex, and more accurate method automatically translates into better business outcomes (Kolassa 2022a). Thus, although AI/ML has been at the center of every forecaster's attention, it will likely not be a silver bullet that will solve all our forecasting problems (Kolassa 2020).

One thing to keep in mind is that computer scientists have primarily driven the development of AI/ML. This poses a problem insofar as forecasting has traditionally been a domain of statisticians and econometricians. One consequence is that AI/ML experts are often unaware of the pitfalls inherent in forecasting in contrast to "traditional" AI/ML use cases, from benchmarking against simple methods to evaluation strategies and adequate evaluation datasets (Hewamalage, Ackermann, and Bergmeir 2022).

In closing, there is one "human factor" difference between the two approaches to forecasting: the traditional methods are more statistical. They are less in vogue than AI/ML. More people are excited about DL and Boosting than about ARIMA and Exponential Smoothing. AI/ML methods are thus far easier to sell to decision-makers, and it may be easier to find people trained in the newer methods than in the conventional ones. Data Scientists trained in AI/ML tend to require a higher salary than statisticians, econometricians, and business analysts trained in the more conventional methods.

Key takeaways

1. In recent years, Artificial Intelligence (AI) and Machine Learning (ML) methods have captured our attention in forecasting. Examples include Neural Networks, Deep Learning, tree-based methods, Random Forests, and various flavors of Boosting.
2. AI/ML methods have outperformed "classical" approaches in terms of accuracy for aggregate time series, but performance comparisons at disaggregate and lower levels of time series are much less apparent.
3. AI/ML methods are non-linear and more complex. They are thus more demanding in terms of data, computing resources, and expertise and need help understanding and debugging.
4. Whether the improved business outcome through improved accuracy is worth the added complexity and higher Total Cost of Ownership of AI/ML methods merits careful analysis.

15

Long, multiple and non-periodic seasonal cycles

Previous chapters discussed incorporating simple forms of seasonality – e.g., the yearly seasonality for quarterly or monthly data – into our models. The present chapter will consider challenges posed by "more complicated" seasonality. The workhorses of forecasting, i.e., Exponential Smoothing and ARIMA, were developed when time series were primarily monthly or quarterly. Today, we collect time series data at a far higher frequency, e.g., daily or hourly. This increased granularity can yield more complicated seasonal patterns. Exponential Smoothing and ARIMA models may not be able to deal with such seasonality.

15.1 Introduction

The first thing to be aware of when we examine more complicated seasonality is that when we discuss seasonality, we always have to know *two* pieces of time-related information: the *temporal granularity* of our time buckets, and the *cycle length* at which patterns repeat seasonally. For instance, we could have a daily granularity and a weekly cycle, meaning that demands on Mondays are similar to each other, also on Tuesdays and so on. However, we might also have hourly granularity and a daily cycle, in which case the hourly demand at 9:00 a.m. is similar between different days. We see that "daily" here crops up in two different places: either as the granularity, or as the cycle length. A term like "daily seasonality" is ambiguous: does it refer to sub-daily time buckets (e.g., hourly) with a daily cycle, or to daily buckets with a cycle that comprises multiple days (e.g., a weekly, monthly, or yearly cycle)? "Monthly seasonality" could similarly refer to daily data with patterns that repeat every month (like payday effects), or monthly data with recurring patterns every 3 months (quarterly cycles) or every 12 months (yearly cycles).

Often, what is meant is clear from the context – if our time series are *all* on daily granularity, then "monthly seasonality" probably refers to patterns of demand that recur every 28-31 days. However, especially if we are talking to someone with a different background, where misunderstandings can easily

TABLE 15.1
Granularities and cycles for seasonality

Granularity	Cycle (buckets per cycle)	Examples
Quarter hour	Day (96 buckets)	Call center demand Electricity demand
Hour	Day (24 buckets)	Traffic flow Residential water demand
Day	Week (7 buckets)	Retail demand
Day	2 weeks (14 buckets)	US paycheck effects
Day	Month (28-31 buckets)	German paycheck effects
Day Week Month	Year (about 365 buckets) Year (about 52 buckets) Year (12 buckets)	Gardening supplies Ice cream demand

happen, it makes sense to be as clear as possible and explicitly say what we are thinking of: "daily data with monthly seasonality." Table 15.1 gives a few examples of possible granularity-cycle combinations.

Figure 15.1 summarizes the different ways in which seasonality can become more complicated than previously discussed:

- *Long seasonality* occurs if a seasonal cycle involves many time buckets. For instance, consider the call volume on a weekday per 15-minute time bucket at a call center that works 24 hours. We can divide each weekday into $4 \times 24 = 96$ periods per daily cycle. If the call center also works on weekends, we will likely have *multiple seasonalities*; see below. We discuss ways to address long seasonality in Section 15.2.

  The term "long seasonality" refers to having many time buckets within a seasonal cycle and does not imply having a long data history. A decade-long time series of quarterly data (4 periods per annual cycle) is not long seasonality. Hourly data with daily seasonality (24 hours a day) border on long seasonality, and longer seasonal periods are clearly so, like 28 to 31 days a month, 96 quarter-hourly buckets in a day, or 168 hours in a week.

- If our demand series exhibits layers of seasonal cycles of different lengths, we refer to *multiple seasonalities.* An example might be hourly website traffic, where the daily pattern differs by the day of the week. In other words, we have a daily cycle of length 24 and a weekly cycle of length $7 \times 24 = 168$.

  In this particular example, the seasonalities are *nested* within each other: the longer seasonal cycle (168 hours) contains multiple entire shorter seasonal cycles (7 days of 24 hours each). We can also have *non-nested*

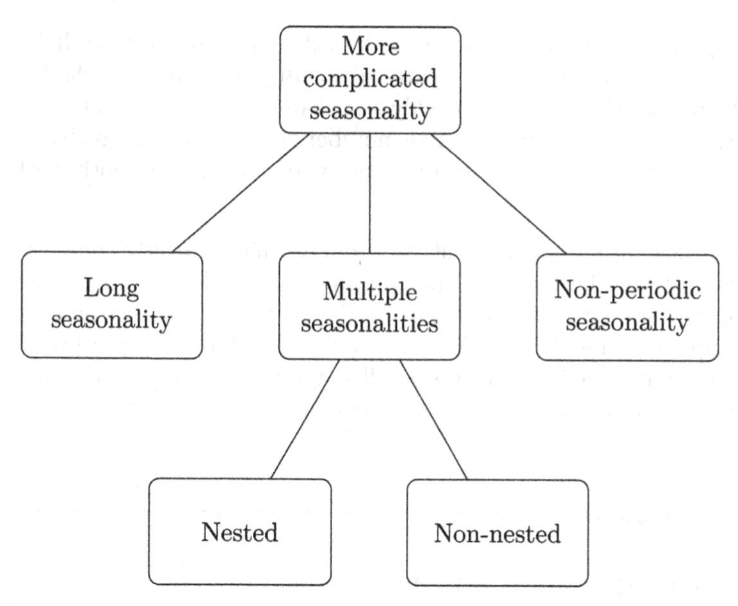

FIGURE 15.1
An overview of more complicated seasonality

multiple seasonalities. For example, daily data may have a weekly (i.e., week of the year) and a monthly pattern. Some weeks fall into two months, so weeks are not nested within months. We discuss multiple seasonalities in Section 15.3 and give a more detailed example in Section 15.4.

Multiple seasonalities usually imply long seasonality but not vice versa. In the example of 15-minute weekday demand at a call center, we have long daily seasonality with 96 periods per cycle. But there is, per se, no other seasonal cycle involved.

- *Non-periodic seasonality* refers to patterns with periodicities that do not align with the standard western calendar or have changing period lengths. Most examples stem from religious or cultural events driven by the lunar calendar, like Easter, Jewish or Islamic holidays, or the Chinese New Year. A more prosaic example is a payday effect if paychecks arrive monthly: the cycle length varies between 28 and 31 days and is thus irregular.

We discuss ways of modeling all three kinds of complicated seasonality in Sections 15.5 and 15.6.

15.2 Long seasonality

Exponential Smoothing and ARIMA models suffer from estimation issues with long seasonality. For instance, Exponential Smoothing must estimate one starting parameter value for each time bucket in a seasonal cycle. It is easy to estimate 12 monthly indices for yearly seasonality, but far more challenging to estimate 168 hourly indices for weekly seasonality. Not only does the numerical estimation take time, but the high number of parameters involved creates model complexity, which can lead to a noisy, unstable, and poorly performing model (see Section 11.7).

Good software design will prevent you from estimating overly complex models by refusing to model long seasonalities in this way. For example, the `ets()` function in the `forecast` package for R will only allow seasonal cycles of lengths up to 24, i.e., hourly data with daily seasonality. If you try to model longer seasonal cycles, the software will suggest using an STL decomposition instead – one of our recommendations below.

15.3 Multiple seasonalities

Data collection happens on ever-finer temporal granularity. This results in time series that exhibit multiple seasonal cycles of different lengths. For instance, hourly demand data can reveal intra-daily and intra-weekly patterns, and hourly supermarket ice cream sales will show intra-daily and intra-yearly seasonality. Cycles can be *nested* within each other (e.g., hours within days within months within years) or are *non-nested* (e.g., weeks within months). They can also change over time. Practical examples with multiple seasonalities include electricity load demand, emergency medicine demand, hospital admissions, call center demand, public transportation demand, traffic flows, pedestrian counts, requests for cash at ATMs, water usage, or website traffic.

Hypothesizing such complicated patterns alone does not always warrant including them in our models. We can examine whether the underlying signal is strong enough to warrant more complex models through (possibly adapted) seasonal plots (see, for example, Figure 15.4). Researchers and practitioners have developed several new (and more complex) approaches to modeling multiple seasonalities. Recall from Section 11.7 that complex models are not *always* better. Don't be surprised if they don't outperform simpler approaches.

15.4 An example: ambulance demand

An example of multiple seasonalities is emergency medical services. To illustrate, we examine a time series of ambulance demand data from a large ambulance service in the United Kingdom. The dataset contains the hourly number of ambulances demanded (or similar incidents). Figure 15.2 shows the daily ambulance demand plotted for 5 years. It is challenging to identify patterns, given the sheer number of data points plotted.

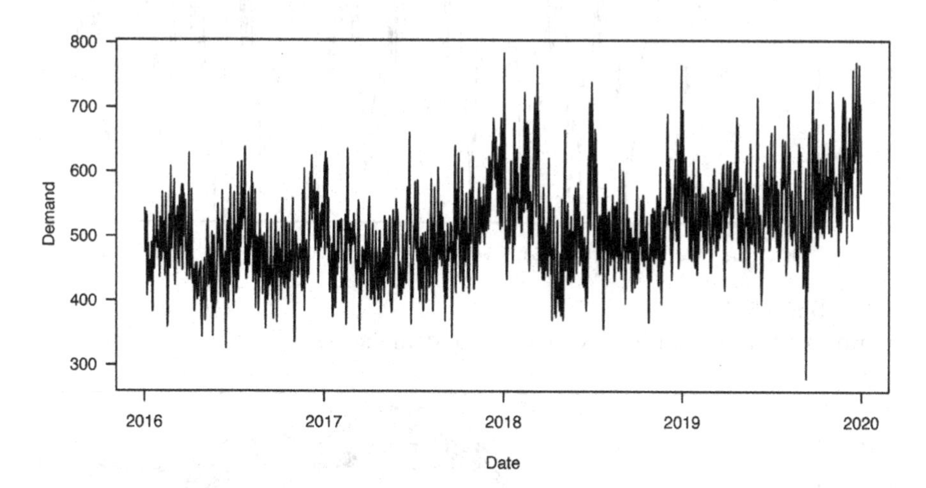

FIGURE 15.2
Daily time series of ambulance demand

Figure 15.3 graphs two weeks of hourly ambulance demand. The regular intra-daily pattern is evident in this plot. The cycles for Monday through Sunday look similar.

Figure 15.4 depicts a seasonal plot of the average number of incidents per hour *separately for each day of the week*, from January 2016 to December 2019. Ambulance demand follows an apparent intra-daily pattern. It decreases between midnight and early morning and then increases until the evening, decreasing again until midnight. The changes in the ambulance demand per hour of the day reflect the population's activities, such as driving, working, hiking, etc. As the number of activities increases during the daytime, we see more people requesting ambulances.

Intra-weekly effects are also visible. People demand more ambulances on Monday and fewer ones on Sunday. This pattern again likely reflects the distribution of human activity throughout the week. Events like weekends, nights out, festivals, rugby matches, etc. shift the place of activities and change the demand for ambulance services. Demand is also higher on Saturday and Sunday between midnight and 5 a.m. This shift also holds for Friday and Saturday between 7 p.m. and midnight. These demand shifts originate from the temporal distribution of human activities and behaviors.

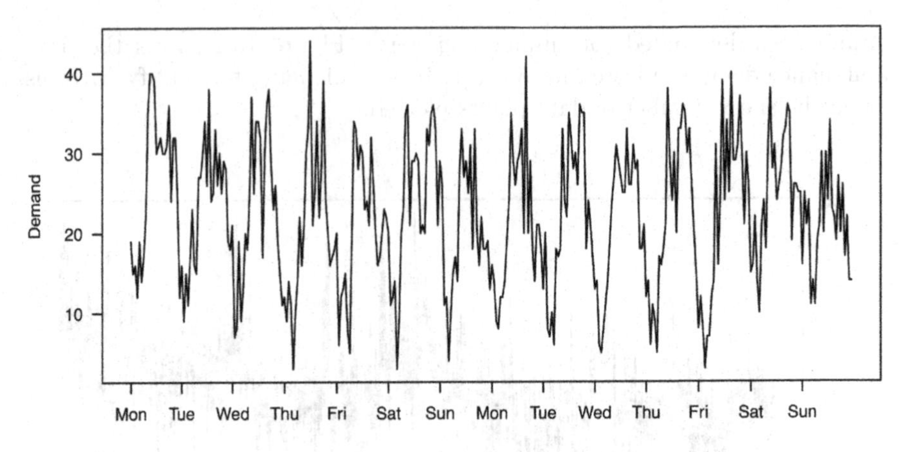

FIGURE 15.3
A two week sample of hourly ambulance demand

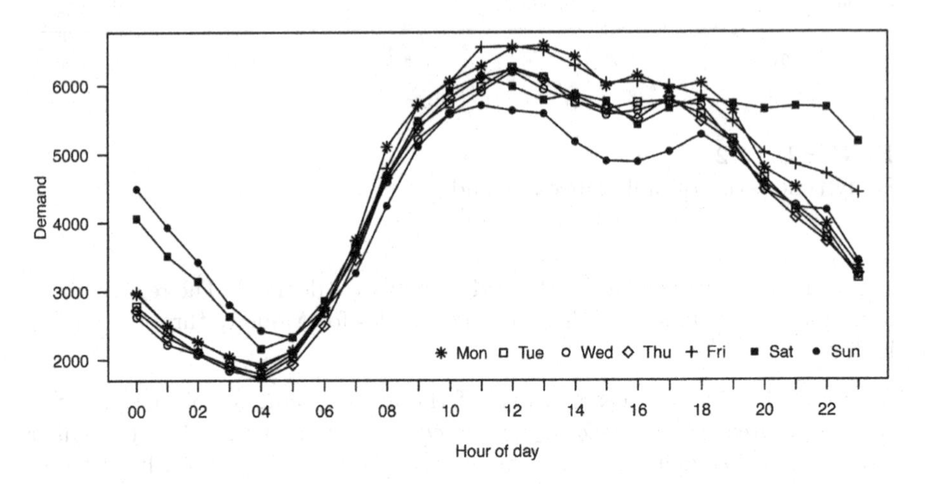

FIGURE 15.4
Average ambulance demand per hour, seperated by the day of the week

15.5 Modeling complicated seasonality

There are four main ways to account for long, multiple, or non-periodic seasonalities in time series forecasting. We can model such seasonality using predictors, lagged terms, seasonal dummy variables, or harmonics.

Predictors

Seasonality has an underlying cause. If we know the drivers of seasonality and if we can measure and forecast them, we can use these measures to model and explain seasonal variation in our time series data. For example, suppose that our firm schedules a product promotion every two weeks. To capture this behavior using seasonality, we would have to deal with the issue of not every month having the same number of weeks. The model would get complicated. But if we can measure whether a promotion occurs, we do not need to model this seasonality; instead, we can simply add this single predictor to your model.

We discuss using predictors in forecasting in Chapter 11. If we can identify predictors that describe or are associated with the seasonal patterns in our time series data, we should include them in the model. We should also investigate each predictor's potential lagged relationship structure (see also Section 11.3).

It can be challenging to measure these predictors over time consistently. Depending on how we use them in our models, we may need to forecast them, reducing their accuracy. For example, if we use temperature as a predictor, we will not know the future temperature. We would need temperature forecasts, which creates an extra layer of uncertainty. If the association between predictors and the forecast variable is weak, or measures are costly to obtain, we should use one of the other methods described below.

Lagged terms

Lagged terms are past values of the variable we want to forecast. Do not confuse these with a lagged predictor, which is the lagged value of a different independent variable. Chapter 10 discusses how to use lags or autoregression to capture the relationship between today's and yesterday's observation. We can model seasonality similarly by using longer lags.

We can rely on this approach to model multiple seasonalities. Suppose we forecast hourly data with intra-daily and intra-weekly seasonality. To model this multiple seasonality, we can include two additional time series predictors into our model: demand during the same hour one day prior and demand during the same hour on the same day of the week one week earlier.

If our time series includes non-periodic seasonality, we can hand-craft a predictor. Suppose we forecast daily data with a payday effect. We can add demand from the same payday of a month one month or multiple months back as a predictor in our model. Careful attention to the lag structures makes "ARIMA-like" models feasible even for long seasonality. We do not need to look at *all* possible lags but pick promising ones based on our domain knowledge of likely seasonal patterns.

TABLE 15.2
An example of daily dummy variables

date	Mon	Tue	Wed	Thu	Fri	Sat
2016-01-04	1	0	0	0	0	0
2016-01-05	0	1	0	0	0	0
2016-01-06	0	0	1	0	0	0
2016-01-07	0	0	0	1	0	0
2016-01-08	0	0	0	0	1	0
2016-01-09	0	0	0	0	0	1
2016-01-10	0	0	0	0	0	0
2016-01-11	1	0	0	0	0	0
2016-01-12	0	1	0	0	0	0
2016-01-13	0	0	1	0	0	0
2016-01-14	0	0	0	1	0	0
2016-01-15	0	0	0	0	1	0
2016-01-16	0	0	0	0	0	1
2016-01-17	0	0	0	0	0	0
2016-01-18	1	0	0	0	0	0

Seasonal dummy variables

An easy way to capture multiple seasonality is to use dummy variables. We previously explained the use of such variables in Section 11.4. A *dummy variable* is a predictor that can take a value of either 0 or 1. A 1 corresponds to *yes*, and a 0 corresponds to *no*. It indicates the presence or absence of a categorical variable potentially influencing demand. We can create dummy variables to represent each seasonal cycle in your data. For example, the ambulance demand data exhibits seasonality at an hourly and weekly level. Adding dummy variables to the data for each of the two seasonal periods will explain some variations in the demand attributable to daily and weekly variations.

Consider our earlier example of ambulance demand forecasting. It would be best to account for the intra-weekly effect when forecasting daily ambulance demand. We can create the dummy variables as shown in Table 15.2. Specifically, six variables represent Monday to Saturday. When the date corresponds to the day of week in each column, we will set that variable to 1, otherwise to 0. Notice that we only require six dummy variables to code seven categories, since the seventh day (i.e., Sunday) occurs when the value of all dummy variables is 0. Accordingly, each dummy variable coefficient can show the level of ambulance demand on each day relative to Sunday, the base period. Similarly, we could create 23 dummy variables to account for hourly seasonality within a day, or 11 dummies to model the months of the year.

While dummy variables are simple to create and interpret, we have to create many such variables when dealing with long seasonality. For example, we would require 167 variables to model the 168 hours a week for our ambulance demand example. This excessive amount of variables can lead to highly complex models. Per Section 11.7, such complexity can yield worse forecasts. Further, dummy variables cannot deal with "fractional" seasonal cycle lengths. For example, there are about 52.18 weeks in a year. We cannot precisely denote the week of the year with dummy variables. For these reasons, forecasters often use harmonics to deal with multiple and long seasonality.

Harmonics

A practical way to model long or multiple seasonalities is to use a small number of *harmonics*, i.e., sine and cosine transformations of time buckets (like dates or hours, whatever the temporal granularity is) with one, two, or three periods per seasonal cycle. These predictors are also called *trigonometric* or *Fourier terms*. Since the sine and the cosine are periodic functions, they are well suited to model periodic behavior. Their flexibility allows using them for fractional seasonal cycle lengths, like the 365.26 days in a year.

For instance, we could transform *dates* using two sine and two cosine waves per *year*, to model intra-yearly seasonality, plus one sine and one coside wave per *week*, to model intra-weekly seasonality. In total, this would yield $2\times2+2\times1=6$ predictors. Figure 15.5 shows the values of all six predictors over one year.

Figure 15.6 shows a sample of hourly ambulance demand at the top with harmonics we could use to model and forecast its multiple seasonalities below. Here, we use one harmonic with one sine and one cosine function each for the intra-weekly and the intra-daily seasonality. We could also use higher order sines and cosines to be more flexible, always keeping in mind that flexibility will not automatically improve accuracy (see Section 11.7).

Harmonics have two key advantages: we can model complicated seasonality with relatively few variables, and the behavior of these variables is smooth. To model the different hours in a week using harmonics, we would need much fewer terms than 167 dummy variables for the 168 hours a week (see below). To illustrate smoothness, consider the following example. Suppose we want to model monthly effects on daily data. We could use 11 dummy variables to model these effects, but using these variables in our model will imply that the forecast jumps rapidly from the last day of one month to the first day of the next – and then stays constant during the entire month, before suddenly jumping again at the beginning of the next month. Such patterns can be real (e.g., with regulatory or tax changes that take effect at the beginning of a month), but most seasonal effects occur more gradually. Figure 15.7 illustrates the difference between monthly dummy variables and harmonics for daily data with intra-yearly seasonality.

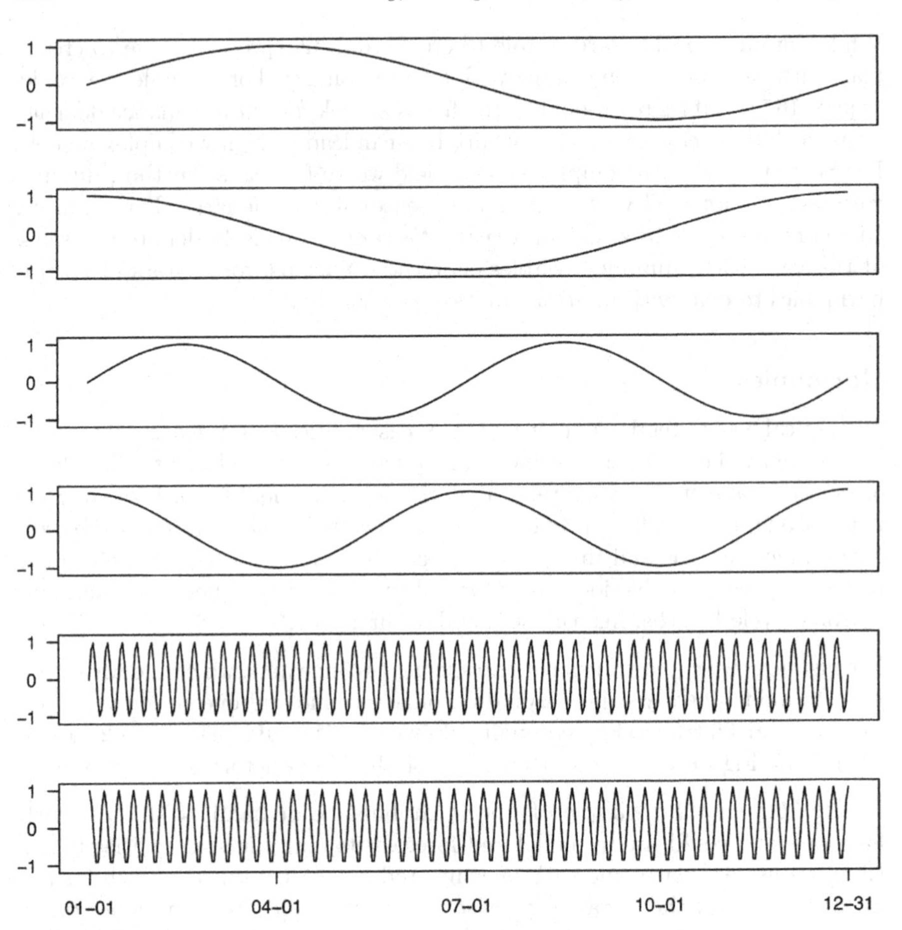

FIGURE 15.5
Predictors to model daily data with both intra-yearly (top four predictors) and intra-weekly (bottom two predictors) seasonality

However, harmonics are less advantageous for modeling "regular" (i.e., short) seasonality. If we have daily data and want to model day-of-week patterns, then six dummy variables as illustrated in Table 15.2 work quite well. Similarly, we can use three quarterly dummy variables to forecast quarterly data or eleven dummy variables for monthly data.

15.6 Forecasting models for complicated seasonality

Table 15.3 lists forecasting models and their capabilities to capture long, multiple or non-periodic seasonalities. Most forecasting models that are appropriate

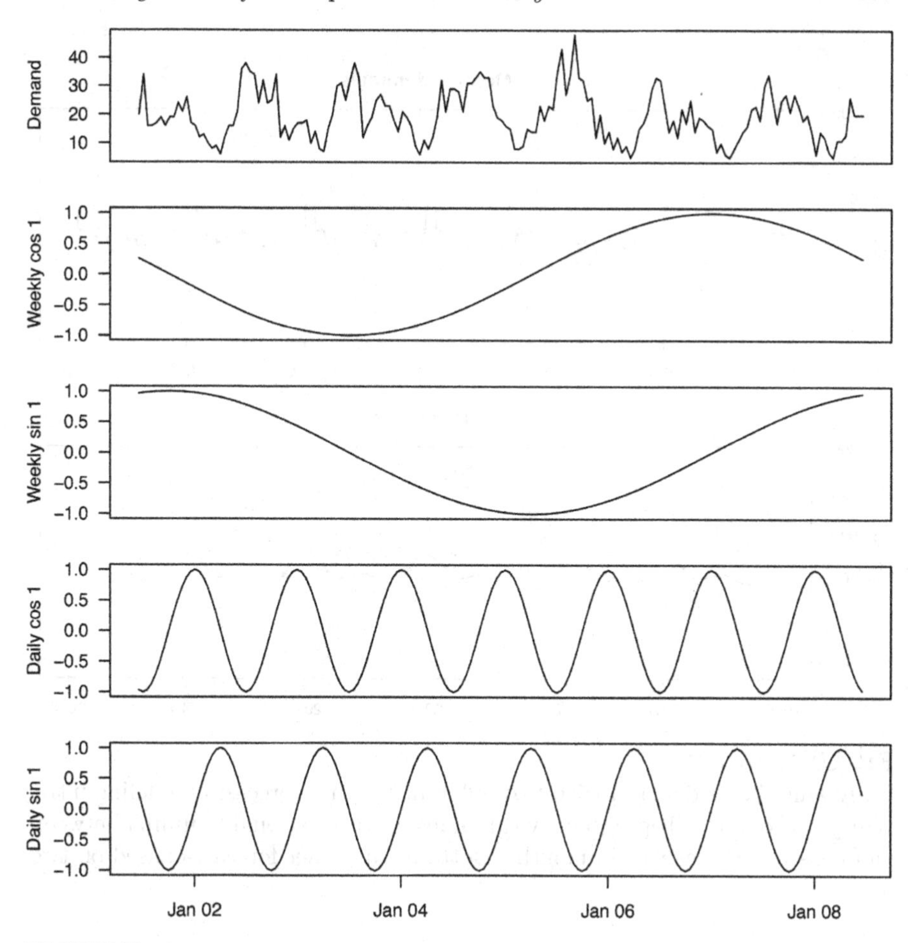

FIGURE 15.6
One week of hourly ambulance demand, with harmonics to model and forecast both intra-weekly and intra-daily seasonality

here use a mixture of lagged terms, seasonal dummies, and harmonics to deal with long and multiple seasonalities, or they can leverage predictors. In the following sections, we briefly review each forecasting model listed in Table 15.3. These models may vary in terms of accuracy, computational time, the ability of terms to evolve over time and flexibility. We refer to flexibility as the ability to incorporate different seasonal terms in the model. You may notice that in Table 15.3, Multiple Linear Regression (MLR) is a flexible model because it could use any of the dummy, harmonic, lagged terms or predictors; however, a model like Double Seasonal Holt-Winters (DSHW) is limited in terms of the types of seasonality that it can include, hence it is less flexible.

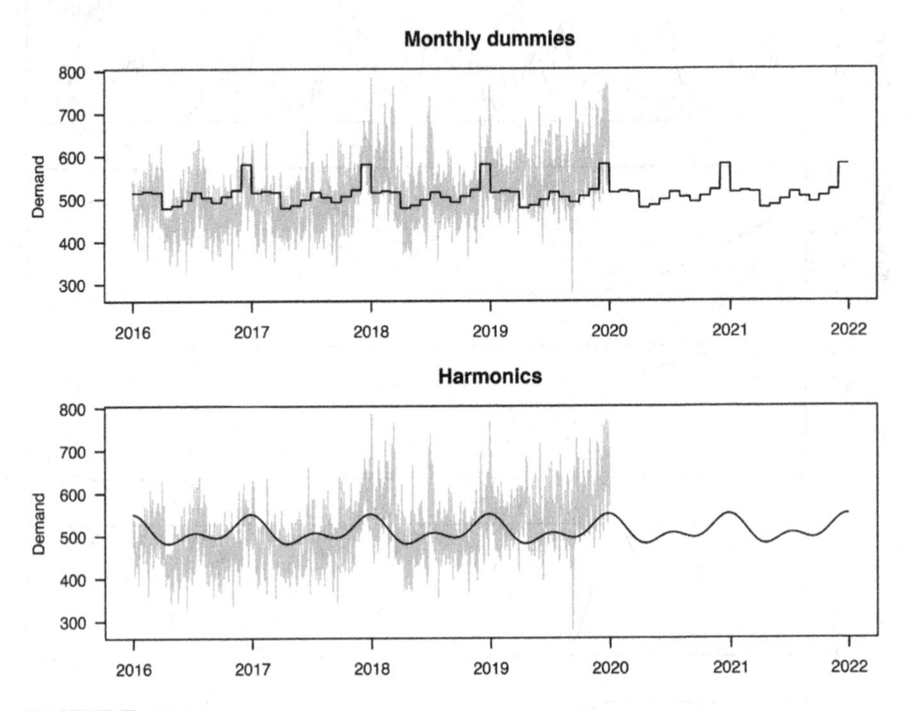

FIGURE 15.7
Daily ambulance demand with two different fits and forecasts modeling intra-yearly seasonality. Top: monthly dummies with questionable jumps between and constant values within months. Bottom: smoother forecasts based on two harmonics

TABLE 15.3
Forecasting models and their link strategies to include multiple seasonality.

Model	Predictors	Lagged Terms	Dummies	Harmonics
Simple average	Depends	Depends	Depends	Depends
MLR	Yes	Yes	Yes	Yes
MSARIMA	Yes	Automatic	Yes	Yes
DSHW	No	No	Automatic	No
TBATS	No	Autoatic	Yes	Automatic
GAM	Yes	Yes	Yes	Yes
Prophet	Yes	Yes	Yes	Automatic
MSTL	No	No	No	No
AI/ML	Yes	Yes	Yes	Yes

Simple averages of seasonal forecasts

A simple way to account for multiple seasonal patterns is to fit different seasonal forecasting models to the data and afterwards average the forecasts. As constituent models, we can use the ones we already know, like Exponential Smoothing or ARIMA for "simple" seasonality, or all the ones below for "more complicated" seasonality.

For example, suppose we have hourly data and suspect seasonal cycles within days, weeks, and years. Then we could fit three different seasonal forecasting models, with period lengths of $m = 24$ (24 hours in a day), $m = 168$ (168 hours in a week) and $m = 8766$ (8766 hours in a year), forecast each model out and take averages of the forecasts. The first model could be Exponential Smoothing or ARIMA, which can still deal with seasonality cycles of length 24, but for the longer weekly and yearly seasonalities, we would need to use one of the models below.

Multiple linear regression

Multiple Linear Regression (MLR) (see also Chapter 11) is a classic method to forecast time series with multiple seasonal cycles. For instance, you could create seasonal dummy variables to encode the day of the week, plus one or more harmonics to account for possible intra-daily and intra-yearly patterns, as in Figure 15.6. We can also model *interactions* between the different seasonal components if we have enough data.

MLR is a fast and flexible approach. It can allow for components to change over time by using weighted regression and weighting more recent observations more heavily. We can also decompose the contributions to seasonality stemming from the various model components (a process that becomes more challenging if we use interactions) for interpretation and debugging. In addition, MLR is available in most programming languages. We can also estimate MLR models quickly.

MLR is implemented in most statistical software packages, from base R to the `scikit-learn` (scikit-learn core developers 2023; Pedregosa et al. 2011) and the `statsmodels` packages (Perktold et al. 2022) for Python.

Multiple/multiplicative seasonal ARIMA

The ARIMA model (see Chapter 10) has been extended to the *Multiple/Multiplicative Seasonal ARIMA (MSARIMA)* model to incorporate multiple seasonalities (J. W. Taylor 2003). MSARIMA allows the inclusion of predictors, and the model components can evolve. However, the model is not very flexible, and the computational cost is very high.

More recently, Svetunkov and Boylan (2020) proposed *Several Seasonalities or State Space ARIMA (SSARIMA)*. SSARIMA constructs ARIMA in a state-space form and allows modeling several seasonalities. While SSARIMA is flexible and allows for including predictors, its computational time to produce forecasts is also very high, especially for high-frequency time series.

R implements both MSARIMA and SSARIMA in the `smooth` package (Svetunkov 2023).

Double seasonal Holt-Winters and variants

Researchers have extended the classical ETS methods (see Chapter 9). J. W. Taylor (2003) introduced a *Double Seasonal Holt-Winters (DSHW)* approach to model daily and weekly seasonal cycles, imposing the same intra-daily cycle for all days of the week. J. W. Taylor (2010) proposed a *triple seasonal model*, adding a separate seasonal state for the intra-daily, intra-weekly, and intra-yearly seasonalities to improve the model proposed by J. W. Taylor (2003). Gould et al. (2008) and J. W. Taylor and Snyder (2012) instead proposed an approach that combines a parsimonious representation of the seasonal states up to a weekly period.

While these models allow for evolving terms and decomposing components, their implementation is not flexible. Moreover, the models do not support the use of additional predictors. Their computational cost is also high.

R implements DSHW in the `forecast` package (Hyndman et al. 2023).

TBATS

An alternative approach to deal with long, multiple and non-periodic seasonalities is *TBATS*. This acronym stands for a method including (optional) trigonometric terms, Box-Cox transformations, ARMA errors, Trend and Seasonal components (De Livera, Hyndman, and Snyder 2011). *TBATS* arranges these components in a completely automated manner. It also allows for terms to evolve.

Some drawbacks of TBATS, however, are that it is not flexible, can be slow to estimate and produce forecasts, and does not allow for predictors in the model. We refer interested readers to Section 12.1 of Hyndman and Athanasopoulos (2021) for more information.

R implements TBATS in the `forecast` package (Hyndman et al. 2023).

Generalized additive models and Prophet

Building on the foundation of the multiple linear regression methodology, *Generalized Additive Models* (*GAMs*) (Hastie and Tibshirani 1990) allow non-linear relationships between the forecast variable and predictors. They can also include harmonics and any lagged terms or predictors in a way similar to MLR.

Facebook introduced *Prophet* (S. J. Taylor and Letham 2018), which utilizes GAMs and accounts for strong seasonal effects through harmonics, piecewise trends, holiday effects (a dummy variable), and additional predictors. The implementation may be less flexible than MLR models, but it is robust to missing values, and structural changes in the trend and can handle outliers well. It is also an automated approach, which makes it easy to use.

The Prophet algorithm is implemented in the `prophet` package for Python (Facebook's Core Data Science Team 2022) and in the `prophet` package for R (S. Taylor and Letham 2021).

Multiple seasonal-trend decomposition using loess

Alternatively, Bandara, Hyndman, and Bergmeir (2021) propose an algorithm to decompose a time series into multiple additive seasonal components analogously to the standard STL decomposition for a single seasonal pattern (see Chapter 7). As in the standard STL decomposition, the different components can be forecasted separately and added together.

R implements the MSTL algorithm in the `forecast` package (Hyndman et al. 2023).

Artificial intelligence and machine learning

In parallel to the above developments, AI/ML methods are an alternative for forecasting time series with multiple seasonality. Neural networks, tree-based methods, Boosting, and other approaches described in Chapter 14 can include features for the hour of the day, the day of the week and the time of year (precisely as for multiple linear regression, see above) and model their effects. In principle, they can also detect and forecast *interactions*, like the effect of the hour of the day differing between different days of the week. However, as discussed in Section 14.5, AI/ML methods usually require much more data than adaptations of classical approaches.

Most AI/ML methods are implemented in various packages in Python, see e.g. Géron (2019), and many are also implemented in R.

Key takeaways

1. If seasonal cycles span many periods, traditional methods like Exponential Smoothing and ARIMA struggle.

2. Multiple seasonal patterns may overlap as we collect and store data on ever finer temporal granularities.
3. Cultural or religious events may have a regular impact that is not aligned with the standard Western calendar and may be non-periodic.
4. Whether multiple seasonalities have a significant impact can be analyzed using appropriate variations of seasonal plots.
5. One can use specialized methods or adapt standard causal models to address long, multiple, and complex seasonalities.

16

Human judgment

So far, we have considered mostly automatic methods for forecasting, where the responsibility of the human forecaster was limited to decide on an appropriate model class, and possibly on the specific model within this class. However, while most "real" forecasts *start out* as the outputs from software, they will very often be *judgmentally adjusted*. While these human inputs have their place and can improve on algorithmically generated forecasts, we must be careful: human intervention is not always positive and can make the forecast worse. Adjusting forecasts also takes time and effort. It is important to carefully examine when and how to adjust forecasts. This chapter examines this challenge.

16.1 Cognitive biases

Any visit to the business book section of a local bookstore will reveal plenty of titles that emphasize that managers need to trust their gut feeling and follow their instincts (e.g., Robinson 2006). This emphasis on intuition illustrates that human judgment is ubiquitous in organizational decision-making.

However, the last decades of academic research have also produced a counter-movement to this view, centered on studying *cognitive biases* (Kahneman, Lovallo, and Sibony 2011; Kahneman 2012). This line of thought sees human judgment as inherently fallible. Intuition, as a decision-making system, has evolved to help us quickly make sense of the surrounding world. Its purpose is not to process all available data and carefully weigh alternatives. Managers can easily fall into the trap of trusting their initial feeling and biasing decisions rather than carefully deliberating and reviewing all available data and options. A key to enabling proper human judgment in forecasting is to reflect upon initial impressions and let further reasoning and information possibly overturn one's initial gut feeling (Moritz, Siemsen, and Kremer 2014).

The presence of cognitive biases in time series forecasting is well documented. We will review several particularly salient ones in this chapter.

Recent data strongly influences forecasters (*recency bias*). They neglect to interpret newer data in the context of the whole time series that has created it. This behavioral pattern is also called *system neglect* (Kremer, Moritz, and Siemsen 2011). It implies that forecasters tend to over-react to short-term shocks in the market and under-react to fundamental and massive long-term shifts.

Further, forecasters are easily misled by visualized data to "find" illusionary patterns in the data that objectively do not exist (*pareidolia*). Simple random walks (such as stock market data) will likely create sequences of observations that consistently increase or decrease over time by pure chance. This mirage of a trend is quickly seen as accurate, even if the past series provides little indication that actual trends exist in the data. Using such illusionary trends for predicting demand can be highly misleading.

If real trends exist in the data, human decision-makers tend to *dampen* them as they extrapolate in the future; that is, their longer-range forecasts tend to believe that these trends are temporary and will naturally weaken over time (Lawrence and Makridakis 1989). Such behavior may benefit long-term forecasts where trends usually require dampening. However, it can reduce accuracy for more short-term forecasts.

This discussion allows us to point out *representativeness* as another essential judgment bias. Forecasters tend to believe that the series of forecasts they produce should look like the series of demands they observe. Consider series 1 in Figure 7.3. As mentioned earlier, the best forecast for this series is a long-run average of observed demand. Thus, plotting forecasts for multiple future periods would result in a flat line. The forecast remains the same from month to month. Comparing the actual demand series with this series of forecasts reveals an odd picture. The demand series shows considerable variation, whereas the forecasts are a straight line. Human decision-makers tend to perceive this as strange and thus introduce variation into their sequence of forecasts, so their forecasts more resemble the series of actual demands (Harvey, Ewart, and West 1997). Such behavior can be quite detrimental to forecasting performance.

Another critical set of biases relates to how people deal with forecast uncertainty. One key finding in this context is *over-precision*: human forecasters tend to under-estimate forecast uncertainty (Mannes and Moore 2013). This bias likely stems from a tendency to ignore or discount extreme cases. The result is that prediction intervals based on human judgment tend to be too narrow – people feel too precise about their predictions. While this bias has been persistent and difficult to remove, recent research has provided some promising results: we can reduce over-precision by forcing decision-makers to assign probabilities to extreme outcomes (Haran, Moore, and Morewedge 2010).

A related bias is the so-called *hindsight bias*: decision-makers tend to believe ex-post that their forecasts are more accurate than they are (Biais and Weber 2009). This bias highlights the importance of constantly calculating, communicating, and learning from the accuracy of past judgmental forecasts.

In demand forecasting and inventory planning, a particular bias exists in *service level anchoring* (Fahimnia et al. 2022). *Anchoring* refers to the unconscious mental process of latching on to a specific number when forming a judgment. Forecasters preparing a point forecast should focus on the most likely outcome or about the 50th percentile of the demand distribution. A firm's service level, often widely known and usually much higher than the 50th percentile, may anchor them on a much higher part of the demand distribution, leading to persistent over-forecasting.

Humans are prone to misallocating their scarce attentional resources on topics that are *salient*, which may not be the ones that are important. This effect has been called *bikeshedding*, after the observation that people will spend proportionally more time discussing a new bike shed costing $50,000 than a new factory costing $10,000,000. As forecasters, we need to be careful not to spend too much time on optimizing less important forecasts (see Section 17.5 on business value).

Statistical models are often difficult to understand for human decision-makers (they are *black boxes*); thus, they trust the method less and are more likely to discount it. In laboratory experiments, the users of forecasting software were more likely to accept the forecast from the software if they could select the model from several alternatives (Lawrence, Goodwin, and Fildes 2002). They tend to be more likely to discount a forecast if the source of the forecast is a statistical model as opposed to a human decision-maker (Önkal et al. 2009).

Quite interestingly, human decision-makers tend to forgive other experts who make errors in forecasts but quickly lose their trust in an algorithm that makes similar prediction errors, which has been called *algorithm aversion* (Dietvorst, Simmons, and Massey 2015). Due to the noise inherent in time series, both humans and algorithms will make prediction errors. Over time, this may imply that algorithms will be less trusted than humans. However, if we think about it, whether the user understands and trusts a statistical model does not, by itself, mean that the model yields bad forecasts! Since we care about our forecasts' accuracy, not our model's popularity, black box or algorithm aversion arguments should not influence us against a statistical model.

In summary, while cognitive biases relate more generally to organizational decision-making, they are very relevant in our context of demand forecasting. However, judgment *can* provide value. Statistical algorithms do not know what they do not know, and forecasters may have domain-specific knowledge that enables better forecasting performance than any algorithm can achieve. A sound forecasting system needs to allow human input to capture this domain-specific knowledge. Recent research shows that, with the proper decision-making avenues, the forecasting performance of human judgment can be extraordinary (Tetlock and Gardner 2015).

16.2 Domain-specific knowledge

One crucial reason human judgment is still prevalent in forecasting is the role of domain-specific knowledge (Lawrence, O'Connor, and Edmundson 2000). Human forecasters may have information about the market that is not (or only imperfectly) incorporated into their current forecasting models. Such domain-specific knowledge, in turn, enables humans to create better forecasts than any statistical forecasting model could accomplish. From this perspective, the underlying forecasting models appear incomplete if they do not include key variables that influence demand. For example, forecasters often note that their models do not adequately factor in promotions, so their judgment is necessary to adjust any statistical model.

However, in times of widespread business analytics, such arguments seem increasingly outdated. Promotions are quantifiable in terms of discount, length, advertisement, etc. Good statistical models to incorporate promotions into sales forecasts are available, as discussed in Chapter 11 (see also Fildes, Ma, and Kolassa 2022).

Besides knowing variables that a forecasting model misses, forecasters may have information that is hard to quantify or codify, that is, a highly tacit and personal understanding of the market. Salespeople may, for example, be able to subjectively assess and interpret the mood of their customers during their interactions. They may also get an idea of the customers' estimate of their business development, even if no formal forecast information is shared. The presence of such information hints at model incompleteness as well. Yet, unlike promotions, some of this information may be highly subjective and difficult to quantify and include in any forecasting model.

Another argument for human judgment in forecasting is that such judgment can identify interactions among predictor variables (Seifert et al. 2015). An interaction effect means that the impact of one particular variable on demand depends on the presence (or absence) of another variable. While human judgment is quite good at discerning such interaction effects, identifying the proper interactions can be daunting for any statistical model due to the underlying dimensionality. For example, the potential number of two-way interactions among ten variables is 45; the possible number of three-way interactions among ten variables is 120. Including many interaction terms in a regression equation can make estimating and interpreting any statistical model challenging (see Section 11.7). Human judgment may be able to pre-select meaningful interactions more readily.

16.3 Political and incentive aspects

Dividing firms into functional silos is often a necessary aspect of organizational design to achieve focus in decision-making. Such divisions usually go hand-in-hand with incentives – marketing and sales employees may, for example, be paid a bonus depending on the realized sales of a product. In contrast, operations employees receive incentives based on cost savings. The precise key performance indicators used vary significantly from firm to firm. While such incentives may provide an impetus for action and effort within the corresponding functions, they create diverging objectives. Such goal misalignment is particularly troublesome for cross-functional processes such as forecasting and sales and operations planning.

Since the forecast is crucial for many organizational decisions, decision-makers try to influence it to achieve their corporate objectives and personal goals. Relying on a statistical model will reduce or eliminate the ability to influence decision-making through the forecast; any organizational influence on the forecast will be visible and will encounter resistance.

Mello (2009) describes seven ways forecasters can be influenced by politics and incentives:

- *Enforcing* behavior occurs when forecasters try to maintain a higher forecast than anticipated to reduce the discrepancy between forecasts and company financial goals. Suppose senior management creates a climate where targets must be met without question. In that case, forecasters may acquiesce and adjust their forecasts accordingly to reduce any dissonance between their forecasts and these targets.
- Relatedly, the game of *spinning* occurs if lower-level employees or managers deliberately alter (usually increasing) the forecast to influence higher-level managers' responses. Such behavior can be a result of higher-level management killing the messenger. If forecasters are criticized for delivering low forecasts, they will adjust their behavior to deliver more pleasant forecasts.
- *Filtering* occurs when forecasters lower their forecasts to reflect supply or capacity limitations. This phenomenon often occurs if operations personnel drive forecasts to mask their inability to meet predicted demand.
- If sales personnel strongly influence forecasts, *hedging* can occur, where forecasts over-estimate demand to move operations to make more product available. Similarly, suppose downstream supply chain partners influence the forecast. In that case, they may inflate demand estimates in anticipation of a supply shortage, wanting to secure a larger proportion of the resulting allocation.

- On the other hand, *sandbagging* involves lowering the sales forecast so that actual demand is likely to exceed it; this strategy becomes prevalent if your organization does not sufficiently differentiate between forecasts and sales and if salespeople's targets are set based on forecasts: if a salesperson can get forecasts to be lower, their sales targets will be lower, and they will be more likely to exceed them and get a big bonus.
- *Second guessing* occurs when influential individuals in the forecasting process override the forecast with their judgment. Such behavior is often a symptom of general mistrust in the forecast.
- Finally, *withholding* occurs when members of the organization fail to share critical information related to the forecast. This behavior is often a deliberate ploy to create uncertainty about demand among other organization members.

In summary, human judgment influences forecasts for good and bad reasons. The question of whether it improves forecasting or not is ultimately an empirical one. In practice, most companies will use a statistical forecast as a basis for their discussion but often adjust this forecast based on the consensus of the people involved.

Forecasting is ultimately a statement about reality; thus, the quality of a forecast can be judged ex-post (see Chapter 17). One can compare whether the adjustments made in this consensus adjustment process improved or decreased the accuracy of the initial statistical forecast in a Forecast Value Added (FVA) analysis (Gilliland 2013). In a study of over 60,000 forecasts across four different companies, such a comparison revealed that, on average, judgmental adjustments to the statistical forecast increased accuracy (Fildes et al. 2009). However, a more detailed look also revealed that smaller adjustments (which were also more frequent) tended to reduce accuracy, whereas larger adjustments increased it. One interpretation of this result is that more significant adjustments usually resulted from model incompleteness, that is, promotions and foreseeable events that the statistical model did not consider. The minor adjustments represent the remaining organizational quibbling, influencing behavior and distrust in the forecasting software. One can thus conclude that a sound forecasting process should only be affected by human judgment in exceptional circumstances and with a clear indication that the underlying model is incomplete. Otherwise, organizations should limit the influence of human judgment in the process.

16.4 Correction and combination

If we use judgmental forecasts in addition to statistical forecasts, two types of methods may help us improve the performance of these judgmental forecasts: combination and correction. *Combination* methods combine judgmental forecasts with statistical forecasts mechanically, as described in Section 8.6. In other words, the simple average of two forecasts – whether judgmental or statistical – can outperform either one (Clemen 1989). *Correction* methods, on the other hand, attempt to de-bias a judgmental forecast before use. *Theil's correction method* is one such attempt, following a simple procedure. A forecaster runs a regression between their past forecasts and past demand in the following form:

$$Demand_t = a_0 + a_1 \times Forecast_t + Error_t. \tag{16.1}$$

The forecaster can then use the results from this regression equation to de-bias all forecasts made after this estimation by calculating

$$Corrected\ Forecast_{t+n} = a_0 + a_1 \times Forecast_{t+n}, \tag{16.2}$$

where a_0 and a_1 in Equation (16.2) are the estimated regression intercept and slope parameters from Equation (16.1). There is some evidence that this method works well in de-biasing judgmental forecasts and leads to better performance of such forecasts (Goodwin 2000). However, we should examine whether the sources of bias change over time. For example, the biases human forecasters experience when forecasting a time series for the first time may be very different from the biases they are subject to with more experience in forecasting the series. Thus, initial data may not be valid for estimating Equation (16.1). Further, if forecasters know that their forecasts will be bias-corrected in this fashion, they may show a behavioral response to overcome and outsmart this correction mechanism.

Human-guided learning is a new integration method that has recently been developed and tested (Brau, Aloysius, and Siemsen 2023). The main idea of this method is to let forecasters only indicate that a special event (e.g., promotion) occurs in a period instead of allowing them to adjust the forecast directly. An algorithm in the background estimates the effect of this event and adjusts the forecast accordingly. This method works surprisingly well in a large-scale retail context.

16.5 Forecasting in groups

The essence of forecast combination methods has also been discussed in the so-called *Wisdom of Crowds* literature (Surowiecki 2004). The key observation in this line of research is more than 100 years old: Francis Galton, a British polymath and statistician, famously observed that during bull-weighing competitions at county fairs (where fairgoers would estimate the weight of a bull, with the best estimate winning a prize), individual estimates could be far off the actual weight of the bull, but the average of all estimates was spot on and even outperformed the estimates of experts. In general, estimates provided by groups of individuals tended to be closer to the actual value than estimates provided by individuals.

An academic debate ensued whether this phenomenon was either due to group decision-making, that is, groups being able to identify the more accurate opinions through discussion, or due to statistical aggregation, that is, group decisions representing a consensus that was far from the extreme views within the group, thus canceling out error. Decades of research established that the latter explanation is more likely to apply. Group consensus processes to discuss forecasts can be highly dysfunctional because of underlying group pressure and other behavioral phenomena.

Group decision-making processes that limit the dysfunctionality inherent in group decision-making, such as the *Delphi method* and the *nominal group technique*, exist. Still, the benefits of such methods for decision-making in forecasting compared to simple averaging are unclear. The simple average of opinions seems to work well (Larrick and Soll 2006), a finding that parallels the analogous result for averaging statistical forecasts (the Forecast Combination Puzzle, see Section 8.6). Furthermore, the average of experts in a field is not necessarily better than the average of amateurs (Surowiecki 2004). In other words, decision-makers in forecasting may be well advised to skip the process of group meetings to find a consensus; instead, they should prepare their forecasts independently. These independent forecasts' simple or weighted average can establish the final consensus. This averaging process filters out the random error inherent in human judgment. Therefore, the benefit of team meetings in forecasting may be more related to improved stakeholder management and accountability rather than improved forecast quality.

This principle of aggregating independent opinions to create better forecasts is powerful but counterintuitive. Experts should be better judges, and reaching team consensus should create better decisions. The Wisdom of Crowds argument contradicts some of these beliefs, since it implies that seeking consensus in a group may not lead to better outcomes and that a group of amateurs can beat experts. The latter effect has been recently demonstrated in the context of predictions in the intelligence community (Spiegel 2014; Tetlock and Gardner 2015). As part of the *Good Judgment Project*, random individuals from across the United States have been preparing probability judgments on global events. Their pooled predictions often beat the predictions of trained CIA analysts with access to confidential data. If amateurs can beat trained professionals in a context where such professionals have secret domain knowledge, the power of the Wisdom of Crowds becomes quite apparent. The implication is that for critical forecasts in an organization, having multiple forecasts prepared in parallel (and independently) and then simply taking the average of such forecasts may be a simple yet effective way of increasing judgmental forecasting accuracy.

Key takeaways

1. Human judgment can improve forecasts, especially if humans possess information that is hard to consider within a statistical forecasting method.
2. Humans have not evolved to deal well with judgment under uncertainty. Cognitive biases imply that human intervention will often make forecasts worse. That a statistical method is hard to understand does not mean that a human forecaster can improve the forecast.
3. Incentive structures may reward people for making forecasts worse. People will try to influence the forecast since they are interested in influencing the decisions made based on the forecast.
4. Measure whether and when human judgment improves forecasts. It may make sense to restrict judgmental adjustments to only those contexts where factual information of model incompleteness is present (e.g., the forecast does not factor in promotions, etc.).
5. When relying on human judgment in forecasting, get independent judgments from multiple forecasters and then average these opinions.

Forecasting quality

17

Error measures

So far, we have studied several forecasting algorithms. Most of these can have many variants, depending on whether we include seasonality, trend, or other components. How do we establish whether one algorithm does a better job at forecasting a given time series than another one? We need to benchmark different forecasting methods and find the best-performing ones. We thus need *forecast quality* key performance indices or *error measures*. This chapter examines the most common error measures for point forecasts and also briefly introduces such KPIs for prediction intervals and predictive densities.

Performance measurement is necessary for management; it is hard to set goals and coordinate activities if we do not measure performance. This statement is true for almost any aspect of a business, but particularly so for forecasting. However, it contradicts how society often treats forecasting (Tetlock and Gardner 2015). Pundits on television can make bold statements about the future without their feet being held to the fire; management gurus are praised when one of their predictions has come true, without ever considering their long-run record. A similar culture can exist in organizations – managerial gut judgments are taken as fact without ever establishing whether the forecaster making the judgment has a history of being spot on or mostly off. Even worse, forecasts are often made in ways that it becomes impossible to examine their quality, particularly if the forecast does not include a proper time frame.

The good news is that in demand forecasting, forecasts are usually quantified ("we expect demand for SKU X to be Y") and come within a time frame ("Z months from now"). Such a specificity allows explicitly calculating the error in the forecast and thus making long-run assessments of this error. Nevertheless, deciding whether a demand forecast is "good" or one forecasting algorithm is "better" than another is not entirely straightforward. We will devote this chapter to the topic.

17.1 Bias and accuracy

This section introduces the concepts of *bias* and *accuracy* in forecast error measurement and provides an overview of commonly used metrics. Suppose we have calculated a single point forecast and observe the corresponding demand realization later. We will define the corresponding error as

$$Error = Forecast - Demand. \tag{17.1}$$

For instance, if $Forecast = 10$ and $Demand = 8$, then $Error = 10 - 8 = 2$. This definition has the advantage that over-forecasts (i.e., $Forecast > Demand$) correspond to positive errors, while under-forecasts (i.e., $Forecast < Demand$) correspond to negative errors, which follows everyday intuition.

With a slight rearrangement, this definition means that

$$Demand = Forecast - Error, \tag{17.2}$$

or "actuals equal the model *minus* the error," instead of "*plus*." The "plus" convention is common in statistics and machine learning, where one would define the error as $Error = Demand - Forecast$. Such a definition yields the unintuitive fact that over-forecasts (or under-forecasts) would correspond to negative (or positive) errors.

Our error definition, although common, is not universally accepted in forecasting research and practice, and many forecasters adopt the alternative error definition motivated by statistics and machine learning. Green and Tashman (2008) surveyed practicing forecasters about their favorite error definition. Even our definition of forecast error in Chapter 9 defines the error in its alternate form – mostly because this alternative definition makes Exponential Smoothing easier to explain and understand.

Whichever convention you adopt in your practical forecasting work, the key takeaway is that whenever you discuss errors, you must ensure everyone is using the *same* definition. Note that this challenge does not arise if we are only interested in *absolute* errors.

We cannot judge demand forecasts in their quality unless a sufficient number of forecasts are available. As discussed in Chapter 1, if we examine just a single forecast error, we cannot differentiate between bad luck and a bad forecasting method. Thus, we will not calculate the error of a single forecast but instead the errors of many forecasts made by the same method. Let us assume that we have n forecasts and n corresponding actual demand realizations, giving rise to n errors, that is,

$$\begin{aligned} Error_1 &= Forecast_1 - Demand_1 \\ &\vdots \\ Error_n &= Forecast_n - Demand_n. \end{aligned} \tag{17.3}$$

Our task is to summarize this (potentially vast) number of errors so we can make sense of them. The simplest way of summarizing many errors is to take their average. The *mean error* (ME) is the simple average of errors,

$$\text{ME} = \frac{1}{n}\sum_{i=1}^{n} \mathit{Error}_i. \tag{17.4}$$

This ME is the key metric used to assess *bias* in a forecasting method. It tells us whether a forecast is "on average" on target. If ME $= 0$, then the forecast is on average *unbiased*, but if ME > 0, then we systematically over-forecast demand, and if ME < 0, then we systematically under-forecast demand. In either case, if ME $\neq 0$ and the difference between ME and 0 is sufficiently large, we say that our forecast is *biased*.

While the notion of bias is important, we are often less interested in *bias* than in the *accuracy* of a forecasting method. While bias measures whether, on average, a forecast is on target, accuracy measures how close the forecast is to actual demand, on average. In other words, while the bias examines the mean of the forecast error distribution, the accuracy relates to the spread of the forecast error distribution.

One metric often used to assess accuracy is the *absolute* difference between the point forecast and the demand realization, that is, the *absolute error*:

$$\text{AE} = \mathit{AbsoluteError} = |\mathit{Error}| = |\mathit{Forecast} - \mathit{Demand}|, \tag{17.5}$$

where $|\cdot|$ means that we take the value between the "absolute bars," dropping any plus or minus sign. Thus, the absolute error cannot be negative. For example, if the actual demand is 8, then a forecast of 10 or one of 6 would have the same absolute error of 2.

The *mean absolute error* (MAE) or *mean absolute deviation* (MAD) – both terms are used interchangeably – is simply the average of absolute errors,

$$\text{MAE} = \text{MAD} = \frac{1}{n}\sum_{i=1}^{n} |\mathit{Error}_i|. \tag{17.6}$$

Note here that we need to take absolute values *before* and not *after* summing the errors. For instance, assume that $n = 2$, $\mathit{Error}_1 = 2$, and $\mathit{Error}_2 = -2$. Then

$$\begin{aligned} |\mathit{Error}_1| + |\mathit{Error}_2| &= |2| + |-2| = 2 + 2 = 4 \\ &\neq 0 = |2 + (-2)| = |\mathit{Error}_1 + \mathit{Error}_2|. \end{aligned} \tag{17.7}$$

The MAE tells us whether a forecast is "on average" accurate, that is, whether it is "close to" or "far away from" the actual, without taking the sign of the error into account.

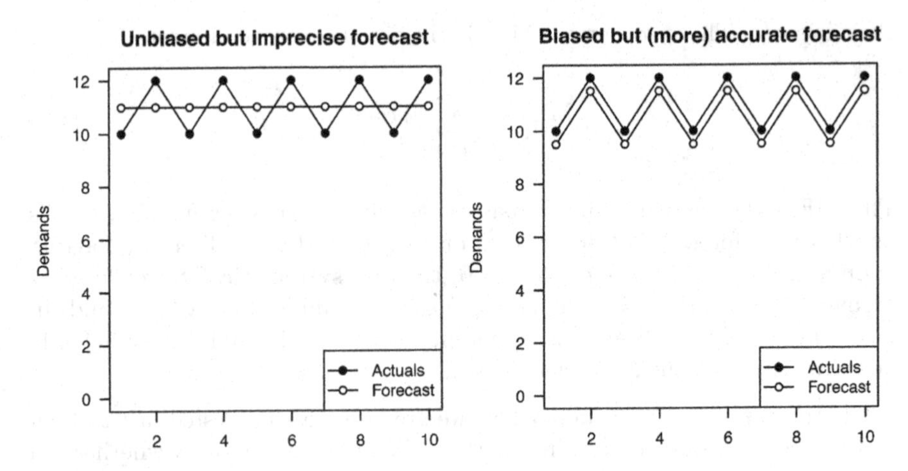

FIGURE 17.1
Bias vs. accuracy

Let us consider an artificial example (see Figure 17.1). Assume that our point forecast is $(11, 11, \ldots, 11)$; that is, we have a constant forecast of 11 for $n = 10$ months. Assume further that the actual observations are $(10, 12, 10, 12, 10, 12, 10, 12, 10, 12)$. Then the errors are $(1, -1, 1, -1, 1, -1, 1, -1, 1, -1)$ and ME $= 0$, that is, our flat forecast is unbiased. However, it is inaccurate since MAD $= 1$. Conversely, assume that the forecast is $(9.5, 11.5, 9.5, 11.5, \ldots, 9.5, 11.5)$. In this case, our errors are $(-0.5, -0.5, \ldots, -0.5)$, and every single forecast is 0.5 units too low. Therefore, ME $= -0.5$, and our forecasts are biased (more precisely, biased *downward*). However, these forecasts are more accurate than the original ones since their MAE $= 0.5$. In other words, even though being unbiased often means a forecast is more accurate, this relationship is not guaranteed. Forecasters sometimes have to decide whether they prefer a biased but more accurate method over an unbiased but less accurate one.

Which of the two forecasts shown in Figure 17.1 is better? We cannot answer this question in isolation. We usually want our forecasts to be unbiased since over- and under-forecasts cancel out in the long run for unbiased forecasts. This logic would favor the first set of forecasts. However, the second set of forecasts better captures the zigzag pattern in the realizations at the expense of bias. To decide which forecast is "better," we would need to assess which leads to better decisions in plans that depend on the forecast, e.g., which forecast yields lower stocks and out-of-stocks (which in turn depends, in addition to point forecasts, on accurate estimates of future residual variances and on other variables, see Section 17.5).

How strong a bias do our error measures need to exhibit to provide evidence that a forecasting method is indeed biased? After all, it is improbable that the average error is precisely equal to zero. To answer this question, we need to standardize the observed average forecast error by the observed variation in forecast errors – much akin to calculating a test statistic. This standardization is the objective of the *tracking signal*, which we can calculate as the cumulative sum of errors divided by the Mean Absolute Error:

$$\text{TS} = \frac{\sum \mathit{Error}_i}{\text{MAE}} \tag{17.8}$$

We constantly monitor the tracking signal. If it falls outside certain boundaries, we deem the forecast biased. A general rule of thumb is that if the tracking signal consistently goes outside the range of ± 4, that is, if the running sum of forecast errors is four times the average absolute deviation, then this constitutes evidence that the forecasting method has become biased.

One other widespread point forecast accuracy measure often used as an alternative to the MAE, works with *squared* errors, that is, Error^2. The square turns every negative number into a positive one, so similarly to absolute deviations, squared errors will never be negative. In order to summarize multiple squared errors $\mathit{Error}_1^2, \ldots, \mathit{Error}_n^2$, one can calculate the *Mean Squared Error* (MSE),

$$\text{MSE} = \frac{1}{n}\sum_{i=1}^{n} \mathit{Error}_i^2. \tag{17.9}$$

The MSE is another measure of accuracy, not bias. In the example in the previous section, the first (constant) forecast yields MSE $= 1$, whereas the second (zigzag) forecast yields MSE $= 0.25$.

Should one use absolute (i.e., MAE) or squared (i.e., MSE) errors to calculate the accuracy of a method? Squared errors have one crucial property: Because of the process of squaring numbers, they emphasize large errors. Indeed, suppose that in the example above, we change a single actual realization from 10 to 20 without changing the forecasts. Then the MEs change slightly to -1 and -1.5, and the MAEs change slightly to 1.8 and 1.5, but the MSEs change drastically to 9 and 11.25.

By squaring errors, the MSE becomes more sensitive to outlier observations than the MAE – which can be a good thing (if outliers are meaningful) or distracting (if you do not want to base your decision-making on outlier observations). If you use the MSE, it is always important to screen forecasts and actuals for large errors and think about what they mean. If these large errors are unimportant in the larger scheme of things, you may want to remove them from the forecast quality calculation or switch to an absolute error measure instead.

In addition, squared errors have one other technical, but fundamental property: Estimating model parameters by minimizing the MSE will always lead to unbiased errors, at least if we understand the underlying distribution well enough. The MAE does not have this property. Optimizing the MAE may lead to systematically biased forecasts, especially when we forecast intermittent or count data – see Section 17.4 and Morlidge (2015) as well as Kolassa (2016a).

Finally, we express squared errors and the MSE in "squared units." If, for example, the forecast and the actual demand are both defined in dollars, the MSE will be denoted in "squared dollars." This scale is somewhat unintuitive. One remedy is to take the square root of the MSE to arrive at the *Root Mean Squared Error* (RMSE) – an error measure similar to a standard deviation and thus somewhat easier to interpret.

17.2 Percentage, scaled and relative errors

All summary measures of error we have considered so far (the ME, MAE/MAD, MSE, and RMSE) have one crucial weakness: They are not scale-independent. If a forecaster tells you that the MAE associated with forecasting a time series with a particular method is 15, you have no idea how good this number is. If the average demand in that series is at about 2,000, an MAE of 15 will imply excellent forecasts! If, however, the average demand in that series is only 30, then an MAE of 15 would be seen as evidence that it is challenging to forecast the series. Thus, without knowing the scale of the series, interpreting any of these measures of bias and accuracy is difficult. One can always use them to compare different methods for the same series (i.e., method 1 has an MAE of 15, and method 2 has an MAE of 10 on the *same* series; thus, method 2 seems to be better), but any comparison between series becomes challenging.

Typically, we will forecast not a single time series but multiple ones, for example, numerous SKUs, possibly in various locations. Each time series will be on a different order of magnitude. One SKU may sell tens of units per month, while another one may sell thousands. In such cases, the forecast errors will typically be on similar orders of magnitude – tens of units for the first SKU and thousands of units for the second SKU. Thus, if we use a point forecast quality measure like the MAE to decide, say, between different possible forecast algorithms applied to all series, our result will be entirely dominated by the performance of the algorithms on the faster-moving SKU. However, the slower-moving one may well be equally or more important. To address this issue, we will try to express all error summaries on a common scale, which we can then meaningfully summarize in turn. We will consider *percentage*, *scaled* and *relative* errors for this.

Percentage errors

Percentage Errors express errors as a fraction of the corresponding actual demand realization to scale forecast errors according to their time series, that is,

$$\text{PE} = \frac{\textit{Error}}{\textit{Demand}} = \frac{\textit{Forecast} - \textit{Demand}}{\textit{Demand}}. \tag{17.10}$$

We usually express these Percentage Errors as percentages instead of fractions. Thus, a forecast of 10 and an actual demand realization of 8 will yield a Percentage Error of $\text{PE} = \frac{10-8}{8} = 0.25$, or 25%.

As in the case of unscaled errors in the previous sub-section, the definition we give for Percentage Errors in Equation (17.10) is the most commonly used, but it is not the only one encountered in practice. Some forecasters prefer to divide the error not by the actual but by the forecast (Green and Tashman 2009). One advantage of this alternative approach is that while the demand can occasionally be zero within the time series (creating a division by zero problem when using demand as a scale), forecasts are less likely to be zero. This modified percentage error otherwise has similar properties as the percentage error defined in Equation (17.10). Note, however, that *how* we deal with zero demands in the PE calculation can have a major impact on what a "good" forecast is (Kolassa 2023a), which becomes especially important if we have many zero demands, i.e., when our time series are intermittent (see Section 17.4 below). The same important point as for "simple" errors applies: all definitions have advantages and disadvantages, and it is most important to agree on a standard error measure in a single organization, so we do not compare apples and oranges.

Percentage Errors $\text{PE}_1 = \frac{\textit{Error}_1}{\textit{Demand}_1}, \ldots, \text{PE}_n = \frac{\textit{Error}_n}{\textit{Demand}_n}$ can be summarized in a similar way as "regular" errors. For instance, the *Mean Percentage Error* is the simple average of the PE_i,

$$\text{MPE} = \frac{1}{n}\sum_{i=1}^{n} \text{PE}_i. \tag{17.11}$$

The MPE is similar to the ME as a "relative" bias measure. Similarly, we can calculate single *Absolute Percentage Errors* (APEs),

$$\begin{aligned} \text{APE} = |\text{PE}| &= \left|\frac{\textit{Error}}{\textit{Demand}}\right| \\ &= \left|\frac{\textit{Forecast} - \textit{Demand}}{\textit{Demand}}\right| = \frac{|\textit{Forecast} - \textit{Demand}|}{\textit{Demand}}, \end{aligned} \tag{17.12}$$

where one assumes *Demand* > 0. APEs can then be summarized by averaging to arrive at the *Mean Absolute Percentage Error* (MAPE),

$$\text{MAPE} = \frac{1}{n}\sum_{i=1}^{n}|\text{PE}_i| = \frac{1}{n}\sum_{i=1}^{n}\text{APE}_i, \tag{17.13}$$

which is an *extremely* common point forecast accuracy measure – but is more dangerous than it looks (Kolassa 2017).

Let us look closer at the definition of percentage errors. First, note that percentage errors are *asymmetric*. If we exchange the forecast and the actual demand, the error switches its sign, but the absolute and squared errors do not change. In contrast, the percentage error changes in a way that depends on the forecast and the demand if we exchange the two. For instance, a forecast of 10 and a demand of 8 yield PE $= 0.25 = 25\%$, but a forecast of 8 and a demand of 10 yield PE $= 0.20 = 20\%$. The absolute error is 2, and the squared error is 4 in either case.

Second, if the demand is zero, then calculating the APE entails a division by zero, which is mathematically undefined; i.e., if the actual realization is zero, then *any* non-zero error is an infinite fraction of it. There are various ways of dealing with the division by zero problem (Kolassa 2023a).

Some forecasting software "deals" with the problem by sweeping it under the rug: in calculating the MAPE, it only sums PE_is whose corresponding actual demands are greater than zero (Hoover 2006). This approach is *not* a good way of addressing the issue. It amounts to positing that we do not care at all about the forecast if the actual demand is zero. If we make production decisions based on the forecast, then it will matter a lot whether our prediction was 100 or 1000 for an actual demand of zero – and such a difference should be reflected in the forecast accuracy measure.

An alternative, which also addresses the asymmetry of percentage errors noted above, is to "symmetrize" the percentage errors by dividing the error not by the actual but by the average of the forecast and the actual (Makridakis 1993), yielding a *Symmetric Percentage Error* (SPE),

$$\text{SPE} = \frac{\textit{Forecast} - \textit{Demand}}{\frac{1}{2}(\textit{Forecast} + \textit{Demand})}, \tag{17.14}$$

and then summarizing the absolute values of these symmetric percentage errors as usual, yielding a *symmetric MAPE* (sMAPE),

$$\text{sMAPE} = \frac{1}{n}\sum_{i=1}^{n}|\text{SPE}_i|. \tag{17.15}$$

Assuming that at least one forecast and the actual demand are positive, the symmetric error is well defined, and calculating the sMAPE poses no mathematical problems. In addition, the symmetric error is symmetric: if we exchange the forecast and the actual demand, then the symmetric error changes its sign, but its absolute value remains unchanged.

Unfortunately, some problems remain with this error definition. While the symmetric error is symmetric under the exchange of forecasts and actuals, it introduces a new kind of asymmetry (Goodwin and Lawton 1999). If the actual demand is 10, forecasts of 9 and 11 (an under- and over-forecast of one unit, respectively) result in APEs of $0.10 = 10\%$. However, a forecast of 9 yields an SPE of $\frac{-1}{9.5} \approx -0.105 = -10.5\%$, whereas a forecast of 11 yields an SPE of $\frac{1}{10.5} \approx 0.095 = 9.5\%$. Generally, an under-forecast by a given difference will yield a larger absolute SPE than an over-forecast by the same amount, whereas the APE will be the same in both cases.

And this asymmetry is not the last of our worries. As noted above, using the SPE instead of the PE means we can mathematically calculate an SPE even when the demand is zero. However, what acually happens for the SPE in this case? Let us calculate:

$$\text{SPE} = \frac{\textit{Forecast} - \textit{Demand}}{\frac{1}{2}(\textit{Forecast} + \textit{Demand})} = \frac{\textit{Forecast} - 0}{\frac{1}{2}(\textit{Forecast} + 0)} = 2 = 200\%. \tag{17.16}$$

Thus, whenever actual demand is zero, the symmetric error SPE contributes 200% to the sMAPE, entirely regardless of the forecast (Boylan and Syntetos 2006). Dealing with zero demand observations this way is not much better than simply disregarding errors whenever actual demand is zero, as above.

Finally, we can calculate a percentage summary of errors differently to address the division by zero problem. We define the *weighted MAPE* (wMAPE) as

$$\text{wMAPE} = \frac{\sum_{i=1}^{n} |\textit{Error}_i|}{\sum_{i=1}^{n} \textit{Demand}_i} = \frac{\text{MAE}}{\textit{Mean Demand}}. \tag{17.17}$$

A simple computation (Kolassa and Schütz 2007) shows that we can interpret the wMAPE as a weighted average of APEs if all demands are positive, where the corresponding demand weights each APE_i. In the wMAPE, a given APE has a higher weight if the related realization is larger, which makes intuitive sense. In addition, the wMAPE is mathematically defined even if some demands are 0, as long as not *all* demands are 0.

Interpreting wMAPE as a weighted average APE, with weights corresponding to actual demands, suggests alternative weighting schemes. After all, while the actual demand is one possible measure of the "importance" of one APE, there are other possibilities for assigning "importance" to an APE, like the cost of an SKU or its margin. As mentioned above, agreeing on a single standard way of calculating the wMAPE within an organization is vital.

TABLE 17.1
The Absolute Percentage Error (APE) if we have only two possible outcomes, low and high, and can also only forecast "low" or "high"

	Low Forecast	High Forecast
Low actual	APE is zero	APE is high
High actual	APE is low	APE is zero

Apart from the problem with dividing by zero, which we can in principle address by using the wMAPE, the MAPE unfortunately has another issue that does not necessarily invalidate its use but that we should keep in mind. The PEs can become very large if the demand is small, even if all demands are above zero. Thus, the APEs for small demands will dominate the MAPE calculation, which incentivizes us to bias our forecast downward.

A helpful way of looking at this is to consider the MAPE as a weighted sum of AEs, with weights that are given as the *reciprocal* of observed demands:

$$\begin{aligned}\text{MAPE} &= \frac{1}{n}\sum_{i=1}^{n}\text{APE}_i = \frac{1}{n}\sum_{i=1}^{n}\frac{|\textit{Forecast}_i - \textit{Demand}_i|}{\textit{Demand}_i} \\ &= \frac{1}{n}\sum_{i=1}^{n}\frac{1}{\textit{Demand}_i}\times|\textit{Forecast}_i - \textit{Demand}_i| \\ &= \frac{1}{n}\sum_{i=1}^{n}\frac{1}{\textit{Demand}_i}\times\text{AE}_i.\end{aligned} \tag{17.18}$$

We see that the *same* AE will get a higher weight in the MAPE calculation if the corresponding demand is smaller. Minimizing the MAPE thus drives us toward reducing the AEs in particular on *smaller* demands – which will usually mean reducing *all* forecasts, i.e., biasing our forecasts downward.

As an illustration, assume our demand has only two possible values, low and high, and that we can also only forecast "low" or "high" (see Table 17.1). If we forecast "low" and demand is low, we have a zero error, and a zero APE. The same holds if we forecast "high" and demand also turns out to be high. Things become more interesting if we forecast wrong. Irrespective of the direction of our error, the AE is the same, i.e., the difference between the high and the low value. What changes between these two possible situations is the weight given to this AE. If we forecast "low" and demand is high, we weight the AE by the reciprocal of the high value, so we get a low APE. But if we forecast "high" and the demand is low, our weight is the reciprocal of a low value, so the APE will be high! Minimizing means forecasting low.

Kolassa and Martin (2011) give a simple illustration of a similar effect that you can try at home. Take any standard six-sided die and forecast its roll.

Assuming the die is not loaded, all six numbers from one to six are equally likely, and the average roll is 3.5. Thus, an unbiased forecast would also be 3.5. What MAPE would we expect from forecasting 3.5 for a series of many die rolls? We can simulate this expected MAPE empirically by rolling a die many times. Alternatively, we can calculate it abstractly by noting that we have one chance in six in rolling a one, with an APE of $|1 - 3.5|/1 = 250\%$, another one-in-six chance of rolling a two, with an APE of $|2 - 3.5|/2 = 75\%$, and so on. It turns out that our expected MAPE is 71% for a forecast of 3.5.

We can use our dice to see what happens if we use a biased forecast of 4 instead of an unbiased forecast of 3.5. Little surprise here – the long-run MAPE of a forecast of 4 is worse than for a forecast of 3.5: it is 81%. However, what happens if our forecast is biased low instead of high? This time, we are in for a surprise: a forecast of 3 yields an expected MAPE of 61%, clearly lower than the MAPE for an unbiased forecast of 3.5. And an even more biased forecast of 2 results in a yet lower long-run MAPE of 52%. Try this with your dice at home!

Explaining this effect requires understanding the asymmetry of MAPE. Any forecast higher than 2 will frequently result in an APE larger than 100%, for example, if we roll a one. Such high APEs pull the average up more than lower APEs can pull it down. The bottom line is that the expected MAPE is minimized by a forecast that is heavily biased downward. Using this KPI can then lead to very dysfunctional incentives in forecasting.

Interestingly, this simple example shows that alternatives to "vanilla" MAPE, such as the sMAPE or the MAPE with the forecast as a denominator, are also minimized by forecasts that differ from the actual long-run average. This asymmetry in the APE creates a perverse incentive to calculate a biased low forecast rather than one that is unbiased but has a chance of exceeding the actual by a factor of 2 or more (resulting in an APE > 100%). These incentives could tempt a statistically savvy forecaster to apply a "fudge factor" to the statistical forecasts obtained using their software, reducing all system-generated forecasts by (say) 10%.

Scaled errors

An alternative to using percentage errors is calculating *Scaled Errors*, where we scale the MAE/MAD, MSE, or RMSE (or any other error measure) by an appropriate amount. One scaled error measure is the *Mean Absolute Scaled Error* (MASE; see Hyndman and Koehler 2006; Hyndman 2006; Franses 2016). Its computation involves not only forecasts and actual realizations, but also *historical* observations used to calculate forecasts, because the scaling factor used is the MAE of the naive forecast in-sample.

Specifically, assume that we have historical observations $y_1, \ldots, y_T$, from which we calculate one-step-ahead, two-step-ahead, and later forecasts

$\hat{y}_{T+1}, \ldots, \hat{y}_{T+h}$, which correspond to actual realizations $y_{T+1}, \ldots, y_{T+h}$. Using this notation, we can write our MAE calculations as follows:

$$\text{MAE} = \frac{|\hat{y}_{T+1} - y_{T+1}| + \cdots + |\hat{y}_{T+h} - y_{T+h}|}{h}. \tag{17.19}$$

To define a scaling factor for this MAE, we calculate the MAE that we would have observed if we had used naive one-step-ahead forecasts in the past. That is, simply using the previous demand observation to forecast the future. The naive one-step forecast for period 2 is the previous demand y_1, for period 3 the previous demand y_2, and so forth. Specifically, we calculate

$$\text{MAE}' = \frac{|y_1 - y_2| + \cdots + |y_{T-1} - y_T|}{T-1}. \tag{17.20}$$

The MASE then is the ratio of MAE and MAE′:

$$\text{MASE} = \frac{\text{MAE}}{\text{MAE}'}. \tag{17.21}$$

The MASE thus scales MAE by MAE′. It expresses whether our "real" forecast error (MAE) is larger than the in-sample naive one-step ahead forecast (MASE > 1) or smaller (MASE < 1). Since the numerator and denominator are on the scale of the original time series, we can compare the MASE between different time series.

Keep two points in mind. First, the MASE is often miscalculated. The correct calculation requires using the *in-sample*, naive forecast for MAE′, that is, basing the calculations on historical data used to estimate the parameters of a forecasting method. Instead, forecasters often use the *out-of-sample*, naive forecast to calculate MAE′, that is, the data to which the forecasting method is applied. This miscalculation also results in a defensible scaled forecast quality measure. Still, it is not "the" MASE as defined in literature (Hyndman and Koehler (2006) give a technical reason for proposing the in-sample MAE as the denominator). As always, one just needs to be consistent in calculating, reporting, and comparing errors in an organization.

Second, as discussed above, a MASE > 1 means our forecasts have a worse MAE than an in-sample, naive, one-step-ahead forecast. This, at first glance, sounds disconcerting. Should we not expect to do better than the naive forecast? However, a MASE > 1 could easily come about using quite sophisticated and competitive forecasting algorithms (e.g., Athanasopoulos et al. 2011 who found MASE $= 1.38$ for monthly, 1.43 for quarterly, and 2.28 for yearly data). For instance, remember that we potentially calculate the MASE numerator's MAE from multi-step-ahead forecasts. In contrast, we calculate the MAE′ in the denominator from one-step-ahead forecasts. It is not surprising that multi-step-ahead forecasts are worse than one-step-ahead (naive) forecasts.

What are the advantages of the MASE compared to the MAPE? First, it is scaled, so the MASE of forecasts for time series on different scales is comparable. This attribute is similar to MAPE. However, the MASE has two critical advantages over MAPE. First, it is defined even when one demand realization is zero. Second, it penalizes AEs for low and high actuals equally, avoiding the problem we encountered in the dice-rolling example. On the other hand, MASE does have the disadvantage of being harder to interpret. A percentage error (as for MAPE) is easier to understand than a scaled error as a multiple of some in-sample forecast (as for MASE).

Relative errors

Yet another variation on the theme of error measures is given by *Relative Errors.* We can calculate relative errors for any underlying error measure, e.g., those we considered above and those we will consider in Section 17.3. They always work relative to some benchmark forecast or forecasting method. Essentially, relative errors answer the question of whether and by how much a given focal forecast is better than some benchmark forecast, as measured by the chosen error measure.

For example, assume we use the naive forecast as the benchmark and decide to use the MSE as an error measure. Assume that the benchmark naive forecast yields an MSE of 10 on a holdout dataset. We now fit our favorite model, calculate forecasts on the *same* holdout data, and again evaluate the MSE of this forecast. Let us assume that this MSE is 8. Then the relative MSE, often abbreviated relMSE ("RMSE" could cause confusion with the *Root* Mean Squared Error) is

$$\text{relMSE} = \frac{\text{MSE}_{\text{focal forecast}}}{\text{MSE}_{\text{benchmark forecast}}} = \frac{8}{10} = 0.8. \tag{17.22}$$

Thus, a relMSE < 1 indicates that our focal forecast performs better than the benchmark, and relMSE > 1 suggests that it performs worse, both in terms of the MSE. Per Chapter 8, it always makes sense to consider simple forecasting methods as benchmarks, and relative errors give us a tool to do so: calculate the relative error of your focal forecast against the historical mean or the naive forecast, using whatever error measure you want.

One disadvantage of relative errors is that they *only* give relative information. Suppose you calculate the relative MAE of a forecast with respect to the naive forecast. In that case, you will learn how much smaller your focal forecast's MAE is than the naive forecast's in relative terms, but you will not know anything in absolute terms. Based on relative measures alone, we can only say whether a focal forecast is better than a benchmark, but not by how much in absolute terms nor how good the benchmark forecast was itself.

17.3 Assessing prediction intervals and predictive distributions

Recall from Section 4.2 that a prediction interval for a given coverage level (e.g., 80%) consists of a lower and an upper quantile forecast of future demand such that we expect a corresponding percentage of 80% of future realizations to fall between the quantile forecasts. How do we assess whether such an interval forecast is any good?

A single interval forecast and a corresponding single demand realization do not yield much information. Even if the prediction interval captures the corresponding interval of the underlying probability distribution perfectly (which is referred to as *perfectly calibrated*), then the prediction interval is expected not to contain the actual realization in one out of every five cases in our example with a target coverage proportion of 80%. If we observe just a few instances, we can learn little about the accuracy of our method of creating prediction intervals. An assessment of calibration requires larger amounts of data.

Furthermore, the successful assessment of prediction intervals requires that we fix the method of creating these intervals over time. Suppose we want to examine whether the prediction intervals provided by a forecaster are 80%; if we allow changing the method of creating prediction intervals between observations, the forecaster could simply set vast intervals for four of these periods (being almost certain to contain the observation) and a very narrow interval for the remaining period (which will probably not contain the observation), creating an 80% "hit rate." Needless to say, such gamed prediction intervals are of little use.

In summary, to assess the calibration of prediction interval forecasts, we will need multiple demand observations from a time period when the method used to create these intervals was fixed. Suppose we have n interval forecasts and that k of them contain the corresponding demand realization. We can then compute the achieved coverage rate $\frac{k}{n}$ and compare it to the target coverage rate q. Our interval forecast looks good if $\frac{k}{n} \approx q$. However, we will usually not exactly have $\frac{k}{n} = q$. Thus, the question arises of how large the difference between $\frac{k}{n}$ and q must be to reasonably conclude that our method of constructing interval forecasts is good or bad. We can use a statistical concept called "Pearson's χ^2 test." We create a so-called *contingency table* by noting how often our interval forecasts covered the realization and how often we would have expected them to do so: see Table 17.2 for this table.

TABLE 17.2
Expected and observed coverage

	Covered	Not Covered
Observed	k	$n-k$
Expected	qn	$(1-q)n$

We next calculate the following *test statistic*:

$$\chi^2 = \frac{(k-qn)^2}{qn} + \frac{\left(n-k-(1-q)n\right)^2}{(1-q)n}. \tag{17.23}$$

The symbol "χ" represents the small Greek letter "chi," and this test is therefore often called a "chi-squared" test. We can then examine whether this calculated value is larger than the critical value of a χ^2 distribution with one degree of freedom for a given α (i.e., statistical significance) level. This critical value is available in standard statistical tables or software, for example, using the `=CHISQ.INV` function in Microsoft Excel. Suppose our calculated value from Equation (17.23) is larger than the critical value. In that case, we have evidence of poor calibration, and we should consider improving our method of calculating prediction intervals.

For example, let us assume we have $n = 100$ interval forecasts aiming at a nominal coverage probability of $q = 95\%$, so we would expect $qn = 95$ of actual realizations to be covered by the corresponding interval forecasts. Let us assume we observe $k = 90$ realizations covered by the interval forecast. Is this difference between observing $k = 90$ and expecting $qn = 95$ covered realizations statistically significant at a standard alpha level of $\alpha = 0.05$? We calculate a test statistic of

$$\chi^2 = \frac{(90-95)^2}{95} + \frac{(10-5)^2}{5} = 0.26 + 5.00 = 5.26. \tag{17.24}$$

The critical value of a χ^2 distribution with 1 degree of freedom for an $\alpha = 0.05$, calculated for example using Microsoft Excel by `=CHISQ.INV(0.95;1)`, is 3.84, which is smaller than our test statistic. We conclude that our actual coverage is statistically significantly smaller than the nominal coverage we had aimed for and thus consider modifying our way of calculating prediction intervals.

In addition, there are more sophisticated (but also more complex) methods of evaluating prediction intervals, like the *interval score* – see Section 2.12.2 in Petropoulos, Apiletti et al. (2022) for a discussion. An alternative is to not assess the prediction interval as such, but to evaluate the two endpoints separately, treating them as two separate quantile forecasts. The method of choice for this is the *pinball loss* (Kolassa 2023b) and its variants. For instance, the M5 competition used a scaled version of the pinball loss to make it comparable across series on different aggregation levels (Makridakis et al. 2022).

Finally, we have discussed that the most informative forecast is a full predictive distribution. How would we evaluate this? The tools for this are so-called *proper scoring rules*, which are unfortunately highly abstract and very hard to interpret. See Section 2.12.4 in Petropoulos, Apiletti et al. (2022) for a discussion and examples.

17.4 Accuracy measures for count data forecasts

Count data, unfortunately, pose particular challenges for forecast quality assessments. Some quality measures investigated so far can be seriously misleading for count data. For instance, the MAE does not work as expected for count data (Morlidge 2015; Kolassa 2016a). The underlying reason is well known in statistics, but you will still find forecasting researchers and practitioners incorrectly measuring the quality of intermittent demand forecasts using the MAE.

What is the problem with MAE and count data? There are two key insights that can help us understand this issue. One is that we want a point forecast quality measure to guide us toward unbiased point forecasts. Put differently, we want any error measure to have a minimum on average if we feed it unbiased forecasts. Unfortunately, the MAE does not conform to this requirement. Whereas the MSE is minimized and the ME is zero in expectation for an unbiased forecast, the MAE is not minimized by an unbiased forecast for count data. Specifically, the forecast that minimizes the expected (mean) absolute error for any distribution is not the expected value of a distribution but its median (Hanley et al. 2001). This fact does not make a difference for a symmetric predictive distribution like the normal distribution, since the mean and the median of a symmetric distribution are identical. However, the distributions we use for count data are hardly ever symmetric, and this deficiency of the MAE thus becomes troubling – and potentially disastrous.

As an example, Figure 17.2 shows three Poisson-distributed demand series with different means (0.05, 0.3, and 0.6), along with probability mass histograms turned sideways. Importantly, in all three cases, the median of the Poisson distribution is zero, meaning that *the point forecast that minimizes the MAE is zero.*

Turning this argument around, suppose we use the MAE to find the "best" forecasting algorithm for several count data series. We find that a flat zero-point forecast minimizes the MAE. This is not surprising after our prior discussion.

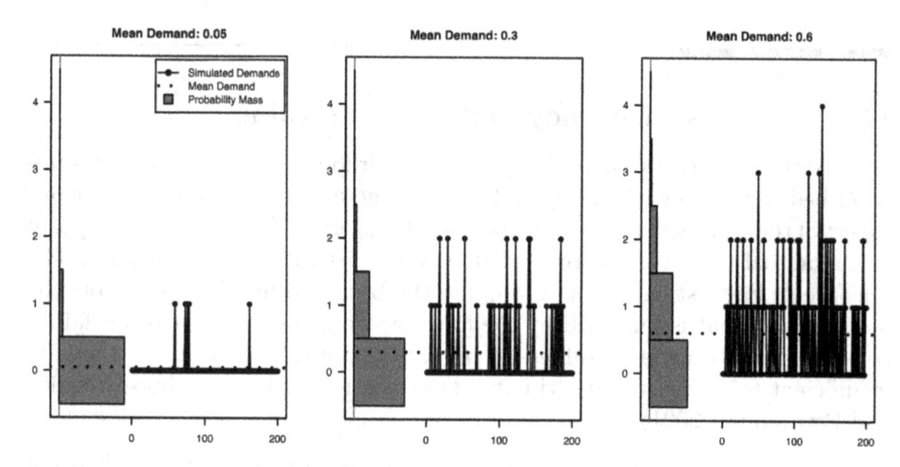

FIGURE 17.2
Poisson-distributed count demand data

However, a flat zero forecast is not useful. The inescapable conclusion is that *we need to be very careful about using the MAE for assessing point forecasts for count data!*

Unfortunately, this realization implies that all point forecast quality measures that are only scaled multiples of the MAE are equally useless for count data. Specifically, this applies to the MASE and the wMAPE.

Finally, the MAPE does not make any sense for intermittent data since the APE is undefined if the actual is zero (Kolassa 2017). There have been many proposals on dealing with this issue, as discussed above. It turns out that how exactly we deal with zero actuals in the context of the MAPE has a significant impact on what the best forecast (in terms of minimizing the expected MAPE) for a given time series is and on the sheer scale of the MAPE we can expect (Kolassa 2023a). Of course, this problem is more prevalent the more intermittent the time series is.

Some quality measures do work as expected for count data. The ME is still a valid measure of bias. However, highly intermittent demands can, simply by chance, have long strings of consecutive zero demands so that any non-zero forecast may look biased. Thus, detecting bias is even harder for count data than for continuous data. Similarly, the MSE retains its property of being minimized by the expectation of future realizations. Therefore, we can still use the MSE to guide us toward unbiased forecasts. However, as discussed above, the MSE can still not be meaningfully compared between time series of different levels, so scaled errors are just as relevant for intermittent series as for non-intermittent ones.

17.5 Forecast accuracy and business value

Forecasters regularly evaluate the quality of their forecasts using error measures described in the previous sections. However, these measures consider just one aspect of the quality of the forecast produced: the error of the forecast compared to the actual outcome. A forecast does not exist for its own purpose. The purpose of forecasting is not to provide the best forecast, but to enable the best decision that is informed by the forecast. Common error measures fail to consider the forecast's usefulness in making better decisions. Thus, they are insufficient to evaluate the ultimate value of a forecasting method (Yardley and Petropoulos 2021).

Although the forecast plays a key role in the decision-making process, it is not the only input. Other elements also have to be considered, often in the form of costs, capacities, constraints, policies and business rules. A number of research projects have demonstrated that the efficiency of the decision-making processes does not relate directly to demand forecasting performance, as measured by standard error measures (Syntetos, Nikolopoulos, and Boylan 2010). The fact that forecasting method A performs better than forecasting method B in terms of error measures (however this is measured) does not necessarily imply that method A will lead to better decisions than B (Robette 2023).

Better forecasts won't earn you money. Better production plans, capacity utilization, or stock control will. The best forecast is not the perfect forecast, or the one with the highest accuracy or lowest error, but the one that allows the best decisions to be made. Forecast accuracy is only a means to an end. And yes, this implies that interpreting forecast accuracy without looking at the larger picture is short-sighted. The mission of forecasters should therefore not be to minimize the error between a forecast and reality, but to maximize the business benefit. (This is why we believe that forecasters also need to understand the larger picture, and to have good communication skills for the necessary cross-functional discussions, see Section 19.1).

When evaluating the quality of a forecast, we thus need to measure the implications of different forecasts. Forecasts generated by method A may be more accurate than forecasts generated by method B, but if subsequent processes mean that they have the same implications (e.g., because we are using forecasts as an input to a production planning process, but logistical constraints mean that both forecasts will lead to the same production plan), then in terms of actual outcomes, both forecasts are equally good.

One way to investigate this is by simulating the processes that turn forecasts into decisions (Kolassa 2023b). Such a simulation is usually not easy, since the following processes are complicated. However, setting up such a simulation, making reasonable simplifications, is often much more enlightening than chasing forecast accuracy improvements without knowing whether they actually translate into business value.

Key takeaways

1. There are numerous forecast accuracy measures.
2. Different accuracy measures measure different things. There is no one "best" accuracy KPI. Consider looking at multiple ones, but be aware that different KPIs reward different forecasts.
3. If you have only a single time series or series on similar scales, use MSE or MAE.
4. Use scaled or percentage errors if you have multiple series at different scales. However, remember that these percentage errors can introduce asymmetries concerning how they penalize over- and underforecasting.
5. Always look at bias. MAE and MAPE can mislead you into biased forecasts, especially for low-volume series.
6. You get what you reward, so choose error metrics that are aligned with the business outcome. Be aware that there is much scope for gaming poorly chosen error metrics, by human forecasters or forecasting tools and models.
7. Forecast accuracy is not necessarily the same as business value. Beware of chasing worthless accuracy improvements.

18

Forecasting competitions

So far, we have considered many different forecasting methods, from purely statistical to judgmental. And we have only scratched the surface in this book; there are many more established forecasting methods. In addition, if you want to implement a forecasting system, each software vendor will have variants on these basic methods and home-grown, possibly proprietary algorithms. How do we decide which of these forecasting methods is the best for our data?

To decide between many different forecasting methods, or indeed between different parameterizations of a single model, we can run a so-called *forecasting competition*, in which various methods compete to provide the best forecast. Such forecasting competitions are not entirely straightforward to run. We therefore devote an entire chapter to describing the generally accepted best way of running them.

18.1 Planning

Similar to a randomized controlled trial in medicine, a forecasting competition needs to be specified and planned (see Figure 2.1). Changing parameters in mid-competition can be problematic. Including additional data after the first data delivery will likely lead to a more significant data cleansing effort than simultaneous delivery. Switching accuracy measures may mean that models change dramatically, possibly invalidating earlier work.

Think about the decision your forecast is supporting. Do you want it for planning marketing and promotions? Or for short-range ordering and replenishment decisions? What you use a forecast for will influence what data you will need. Once you know the purpose of a forecast, the time granularity, location dimensions (see Chapter 13), the forecast horizon, and what accuracy measures to use (see Chapter 17) will become apparent. You may need forecasts for different purposes, requiring you to carefully consider aggregation and forecast hierarchies (again, see Chapter 13).

Plan for an iterated process during the competition. Perhaps the people that perform the forecasting are already familiar with your data. For example, they may belong to an established forecasting group within your business and have been forecasting your time series for years. In that case, they may already understand your data's quirks and how forecasts are created and used. However, you may perform the forecasting competition on data from a new subsidiary, or involve external consultants or third-party forecasting software vendors. In that case, they must understand your data before calculating meaningful forecasts. It is best to introduce the data via live discussions, either face-to-face or via web conferencing. Written data descriptions are helpful, but forecasting experts will always have additional questions. You would not expect your doctor to provide a reliable diagnosis based on a written report of your symptoms either: context, nuance, and undocumented knowledge matter.

As discussed in Chapter 17, numerous forecast accuracy measures exist. Invest some time in choosing good ones to use for comparison. Specific measures may be unusable for your data, like, e.g., MAPE for data with zeros. Or they may lead to biased forecasts, like MAE for low-volume data. If you use multiple series on different scales, make sure your accuracy measures can meaningfully be compared and summarized across scales using percentage or scaled error measures. Think about using more than one error measure – each performance index is better than others at detecting different problems in your forecasts. However, if you do use different accuracy measures, be open to getting and processing different forecasts, one for each error measure, since different forecasts are optimal for different accuracy measures. You should assess forecasts on both bias and accuracy, as described in Chapter 17.

If your forecasting competition includes external parties, do not be afraid of involving them already in the early planning stage. After all, you are planning on trusting these people's software and expertise with a mission-critical process in your business, so you should be able to trust that they know what they are doing. A dedicated forecasting software provider or consultant may know more about forecasting and have more expertise with more different datasets than your in-house forecasting group, although your in-house experts will likely know your specific data better. Tap this external expertise. Discuss your business and your data with the vendor. Consider their suggestions about what data to include in the forecasting competition. Of course, the vendor will look out for themselves first, but that does not mean their proposals will be useless. In addition, this kind of early discussion also allows you to gauge their expertise and commitment. Try to get subject matter experts to participate in these discussions, not just salespeople.

18.2 Data

After you have thought about what you need the forecast for, you can start collecting data. Make the dataset representative. If the forecast competition aims at identifying a method to forecast a small number of highly aggregated time series, perform the competition on a small number of highly aggregated time series. If you are looking for an automated method that can perform well on thousands of series, use an extensive dataset with thousands of series. If you only give out 20 series, the forecasters will tune each model by hand. Such fine-tuning will not be possible in a production environment with thousands of series – so use more data in the competition.

Does your day-to-day data contain outliers or invalid periods? Make sure to include such series. Conversely, do you plan on forecasting only pre-cleaned data? Then clean the data in the same way before you release it to the competition. Do you have causal drivers, like promotions or weather? Then include these effects in the dataset you provide. As in our example in Chapter 8, the causal effects may or may not improve the forecast, but you will not know if you do not try. Make the data you release for the competition representative of the actual forecasting task you need for your business.

However, as Chapter 11 notes, causal models can require *forecasts* of causal drivers. If your business depends heavily on the weather, your demand forecast will need to rely on weather *forecasts*. Suppose you run your forecasting competition using *actual* weather instead of forecasted weather. In that case, you are pretending that you know the future weather perfectly, and your forecasts will be better than if you relied on weather forecasts instead. Your causal models will appear to perform better than in a production environment. Thus, include *forecasts* of your causal drivers for the forecasting period. As a general rule, a forecast prepared in a particular period should only use the information available in that period.

18.3 Procedure

In principle, the procedure of a forecasting competition is simple: collect data, calculate forecasts, and evaluate the forecast. Since we have already discussed data selection and collection, we now focus on the remaining steps.

One key aspect in forecasting competitions is to hold back evaluation data (*hold-out sample*). For instance, you could collect demand data from three years, give out the first two years' demands, keep the third year's demands for evaluation, and require forecasts for this third year. If the forecaster knows the third year's demands, they may succumb to the temptation to "snoop," tweaking their forecasts until they perform best on the *known* evaluation sample. Of course, that will not work in an actual production environment. Avoid the opportunity to cheat, especially for external vendors.

This discussion leads us to a related aspect: why should we use separate evaluation data? Is assessing how well a forecasting method fits the historical

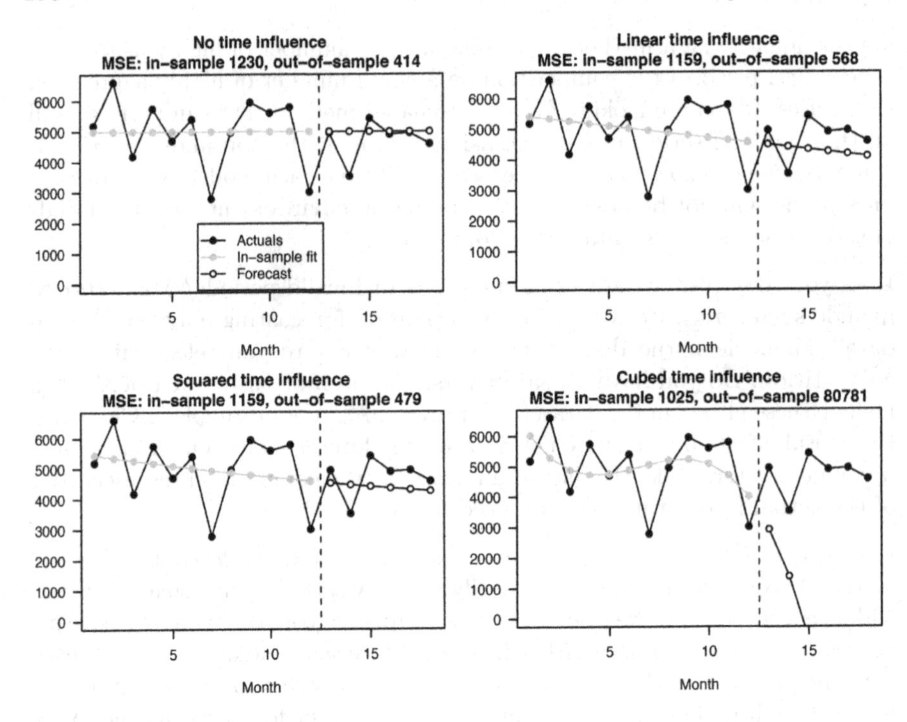

FIGURE 18.1
In-sample and out-of-sample performance for more complex models

data insufficient? If one method performs better in-sample than another method, should it not yield better forecasts, too? Unfortunately, this appealing logic does not work. As a rule of thumb, more complex models deliver better in-sample fits than simpler models (e.g., a seasonal vs. non-seasonal or trend vs. no-trend model). But beyond some optimal complexity, in-sample fit keeps improving while out-of-sample forecast accuracy starts deteriorating. The reason is that the more complex models start fitting to noise instead of capturing a signal.

Figure 18.1 illustrates this effect, using the "hard to forecast" series from Figure 1.1 and giving MSEs in thousands. We fit four models of increasing complexity to the first 12 months of data and forecast the last six months. As we see, the more complex the models are, as reflected in more flexible time influences, the closer the in-sample fit mirrors historical data and the lower the in-sample MSE. But out-of-sample forecasts get worse and worse. Thus, the in-sample fit is not a reliable guide to out-of-sample forecast accuracy, and we should never rely on in-sample accuracy to judge a forecasting method.

Let us continue with the actual procedure of a forecasting competition. You can run either *single origin* or *rolling origin* forecasts. In a single-origin setting, as in Figure 18.1, you might give out 12 months of data and require forecasts for the next six months, allowing you to assess forecasts on horizons between one and six months ahead. In a rolling-origin forecast competition, you would give out 12 months of data and require six months of forecasts. You then provide the actual demand for one more month, requiring five more months of forecasts, etc. In such a rolling origin set<up, the forecast method can learn from each additional month of historical data and adapt. This more closely mimics actual forecasting processes, which repeat and iterate, adapting to new information. Plus, this process gives you more forecasts to evaluate. On the other hand, in a rolling origin setup we need to organize *multiple* exchanges of forecast and actual data and keep careful track of a forecast's "vintage": Was a given forecast for June a one-step-ahead forecast based on data until May or a two-step-ahead forecast based on data through April? Overall, rolling origin forecast competitions are more realistic but require more effort, primarily if you communicate with one or multiple external vendors. If your internal forecasting group runs the competition, rolling origins may simply need a few more lines of code.

Figure 18.2 provides an example of rolling origin forecasting. We use 7 data points as the initial time series (i.e., in-sample) to forecast three subsequent periods that we use for assessment (i.e., out-of-sample). Each iteration adds a new data point at the end of the series and resets the forecasting origin. The procedure stops when the number of periods remaining at the end of the series equals the length of the forecast horizon. We then calculate forecast accuracy for each origin and average the accuracy across all origins.

Finally, examine different angles when evaluating the forecasts of different methods. Summarize forecasts per series (if you have more than one forecast per series), then summarize these sums across all series, e.g., by taking averages of errors. Considering different forecasting horizons, look at how errors behave for one-step, two-step, and longer-range forecasts. Did you include series with outliers (in the historical or the future period) in your dataset? Or series strongly influenced by seasonality or causal factors? If so, check how your forecasts performed on those. Do not only look at averages of errors over series, but also at forecasts that performed terribly. Two methods may yield similar errors on average, but one of the two methods may break down badly in certain circumstances. Such rare spectacular failures can erode users' trust in a forecasting system, even if forecasts are good on average. Assess a method's robustness. Finally, consider discussing results with the forecasters, whether internal or external – you may learn interesting things this way.

18.4 Additional aspects

Of course, the main focus of our forecasting competition lies in forecasting and accuracy. However, as discussed above, forecasting does not happen in a

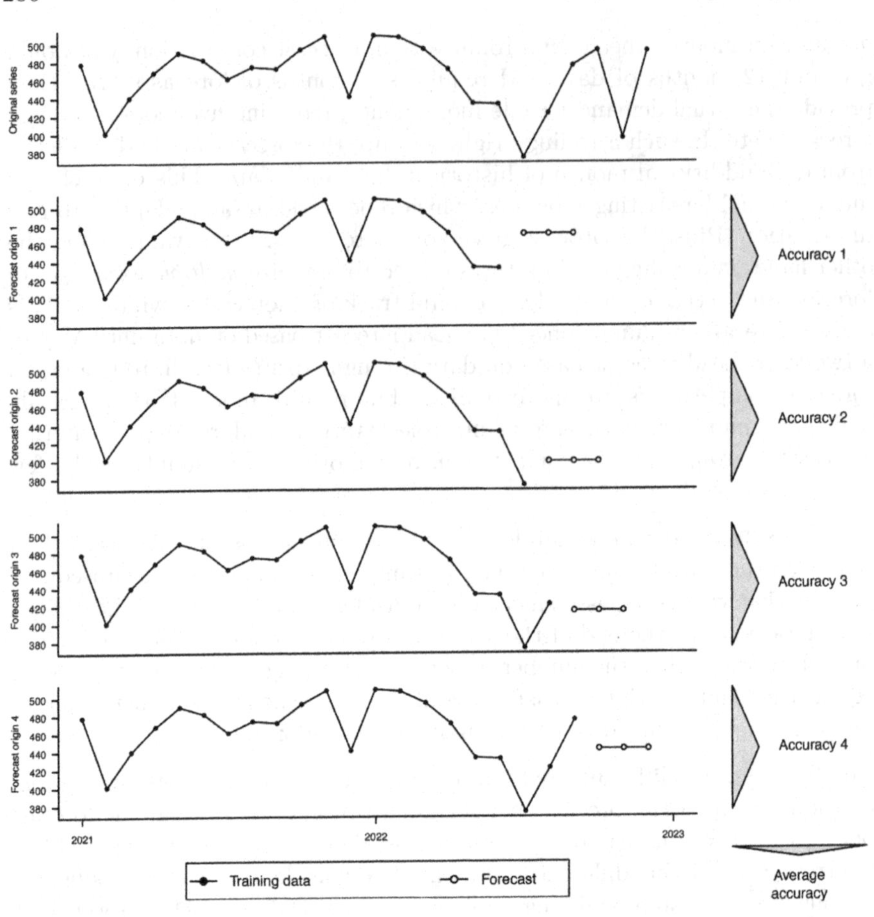

FIGURE 18.2
A rolling origin forecasting competition with an initial length of 18 periods, 4 forecasting origins and a forecast horizon of 3 periods

vacuum. We want forecasts in the first place to support business decisions, like how many units of a product to assemble. It is likely not the forecast itself but the final business decision that influences the bottom line. Good forecasts do not earn or save money – good *decisions* do. And depending on subsequent decision-making processes, forecasts with very different accuracies may yield similar impacts on the bottom line. For instance, if the actual demand is 600 and the two methods generate forecasts of 700 and 900 units, the first method is more accurate. However, if your operations constrain you to assemble in batches of 1,000, both forecasts would likely lead you to the same business decision: building 1,000 units. Key performance indicators, like the service level and the overstock, would have been equivalent. For all practical purposes, spending additional funds to obtain and use the more accurate forecast would have been a waste. (However, if such situations occur frequently, you may want to make your production more flexible.) Thus, it makes sense to simulate the entire process, including forecasting and subsequent decision-making.

Finally, there are other aspects of a forecasting method beyond forecasting accuracy. A method may be highly accurate but require a lot of manual tuning by statistical experts, which can become expensive. Further, more complex forecasting methods can be more challenging to explain and "sell" to other stakeholders. The forecasts from these methods stand a higher chance of being ignored even if they are more accurate (P. F. Taylor and Thomas 1982). Or the method may take too much time to run – one technology may give you all your forecasts within seconds, while the other may need to run overnight. Last but not least, when your forecasting methods must apply to large datasets, slight differences in computational time scale quickly. Using a faster method will not only allow you to save costs, but also to save energy usage, reducing the environmental impact of your forecasting operation – the CO_2 emissions of large data centers are not trivial. Faster forecasting also allows more scenario planning, like changing causal variables and simulating the impact on demands. One forecasting software suite may be stand-alone and may require expensive integration into your ERP system. Another one may already be integrated with your ERP system or have well defined interfaces that connect well with your existing databases.

Key takeaways

1. Plan out your forecasting competition ahead of time.
2. Make sure your competition mirrors your actual forecasting situation regarding data and knowledge and the decisions the forecast must support.
3. Do not be afraid of involving external experts in setting up a forecasting competition. Of course, make sure not to be taken for a ride.
4. *Always* hold back the evaluation data.
5. Evaluate forecasts by slicing the data in different ways to understand a forecasting method's strengths and weaknesses.
6. Forecast accuracy is not everything. Look at the larger picture, from decisions based on the forecast to ease of integration.

Forecasting organization

19

Leading forecasters and forecasting teams

Forecasting is perfomed by humans in teams, and both individual forecasters and forecasting teams need to have certain capabilities to succeed, which we examine in this chapter. We also address how forecasters typically tick, what they require to do a good job, and what you can expect from your forecaster.

19.1 The ideal forecaster

What attributes should you look for when building a world-class forecasting team? What are the essential traits or qualifications forecasters need? What skills should your forecasters bring to the table right from the start, and what skills can they reasonably develop over time?

If you are an aspiring forecaster with an eye to your personal development and career, how do you present yourself to your potential next employer? What are your strengths and your areas of improvement?

Of course, one could have many different requirements for an ideal forecaster. A weather forecaster will have a different job profile than a supply chain forecaster or someone working at a central bank to forecast macroeconomic variables. Nevertheless, we believe that there are four high-level traits that every forecaster should have (see Figure 19.1, adapted from Kolassa 2014): an understanding of *statistics*, *programming* and the *business context*, as well as proficiency in *communication*. Let's look at all of these in turn.

Statistics

"Statistics" means formal statistical training and broader statistical thinking. Concerning the former, a forecaster should understand what a likelihood or a probability distribution represents. Classical time series algorithms like ARIMA (see Chapter 10) are not as valuable for actual forecasting as their ubiquity

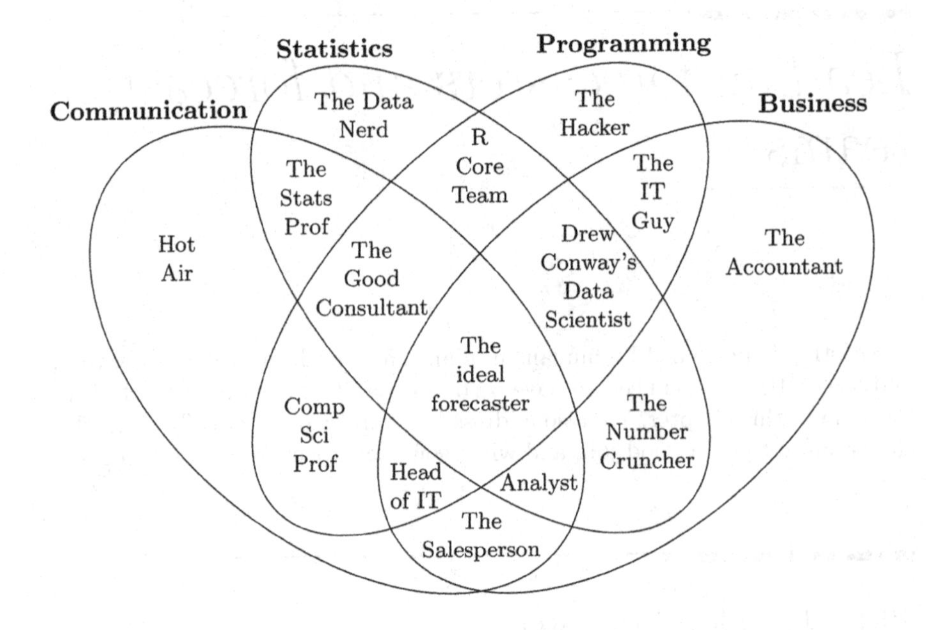

FIGURE 19.1
The ideal forecaster

in forecasting textbooks might suggest, but forecasters should have a basic understanding of these methods.

Concerning broader "statistical thinking," statistics is not a strict set of step-by-step instructions to analyze data. Instead, it is the art and the science of dealing with uncertainty. You think like a statistician if you understand that the data you observe stems from an ambiguous process involving unobserved and unknown dynamics – and so will any future data you are forecasting. Thus, the achievable forecast accuracy is inherently bounded: if you forecast the flip of a fair coin, even the best forecasting algorithm will not get you an accuracy above 50%.

Statisticians – and we are using this term broadly to include data scientists and business analysts that have internalized the statistical way of thinking – understand the influence of noise and uncertainty. The observations we use are noisy. This noise carries over to our fitted forecasting models. And the observations we want to forecast are themselves noisy. Collecting larger and larger datasets does not guarantee absolute accuracy, nor does leveraging ever more predictors (compare Section 11.7 on the bias-variance trade-off, which a forecaster *must* understand). Neither does throwing more money or processing power at the problem guarantee better forecasts.

Programming

Forecasting heavily uses computing, whether on a personal laptop or in the cloud. "Programming" does not mean software development but implies a broader familiarity with scientific computing. The tools a forecaster should be familiar with depend on the environment. Python is (currently) ubiquitous, and newer AI/ML methods typically come in Python frameworks. Some SQL knowledge is essential to be able to deal with databases. R is also often helpful as the language in which academic forecasters and statisticians implement new algorithms and methods. Besides these open-source tools, a forecaster must be familiar with a company's commercial forecasting tools. Given some basic familiarity with computing languages, obtaining such proficiency is usually not too difficult. Even Microsoft Excel can still be surprisingly useful in the hands of an experienced user.

The term "Programming" here also includes "meta-programming" skills such as experience in DevOps, agile methodologies such as Scrum or Kanban, or the ability to write coherent requirements documents.

Business

A forecast is a tool to make a decision, such as capacity planning for a call center, replenishment plans for a supermarket, or even long-term strategic make-or-buy decisions. Forecasting does not happen in isolation. Understanding the business context will frequently be more critical than more profound knowledge of statistics and programming. For example, electricity demand exhibits intra-daily, intra-weekly, and intra-yearly seasonality (see Section 15.3) and depends on weather conditions, especially temperature. Domain-specific causal factors drive call center demand: support for business software will be more needed during working hours, but support for home products will exhibit (relatively) higher demand outside of working hours. Retail demand also exhibits multiple seasonalities and can be highly intermittent at SKU $\times$ store $\times$ day granularity. Forecasting the online channel poses entirely different challenges than forecasting brick-and-mortar demand. Understanding the business context is thus an essential factor in identifying reliable forecasting models and data requirements.

Forecasters do not need to know everything about their specific business domain, but they must be able to *ask the right questions*. A forecaster trained at an electricity exchange may not think about the lumpiness of demand in retail forecasting because it is simply outside their experience. Of course, diverse prior backgrounds may yield creative approaches to solving problems, but every forecaster must have a fundamental understanding of their business.

Communication

Finally, communication skills are indispensable for forecasters beyond statistical, technical, and domain knowledge. They will frequently communicate in very different circumstances and with highly diverse stakeholders. One day, they may need to draft and discuss a requirements document for a software developer. The forecaster may meet with an external forecasting software provider in the afternoon. The next day, they may need to explain an anomalous or disappointing forecast to a business user. The day after, they will need to explain to senior management what steps they plan on taking to make sure a disappointing forecast like this will not happen again – ideally using precisely the right amount of technical/statistical jargon to impress management with their competence, but not so much as to come across as out of touch with day-to-day business realities (compare the "business knowledge" requirement for a forecaster as discussed above).

Forecasters will need to mediate between different viewpoints. Sometimes the "statistically correct" approach would be prohibitively expensive. And conversely, a costly method may still be cost-effective. Business users are often confident that including specific predictors will *absolutely* improve the forecasts. The forecaster may need to explain that this is not necessarily the case (again, compare Section 11.7 on the bias-variance trade-off). Sometimes senior management is enamored with the newest and flashiest machine learning technology (possibly aided and abetted by outside vendors and consultants). The forecaster may need to bring them back to earth gently. The ability to connect emotionally to stakeholders, communicate effectively and with empathy, and understand tailoring their communication approach to the audience will ensure that the forecasting function works well and is respected within the business.

19.2 How do forecasters tick?

One can enter a forecasting function through three main career tracks, and your forecaster's background will influence their mindset. The most common backgrounds are:

- Information Technology (IT)
- Business
- Data Science or Business Analytics

A few years back, before the advent of modern Data Science and Business Analytics, most forecasters came from a backgrund in either IT or Business. Many, if not most, of these forecasters were self-trained. This state of affairs has recently changed, with an explosion of educational possibilities in Data

Science and Business Analytics. Nowadays, if you hire a college graduate for a forecasting position, your applicants often come with such a background. And any applicant shifting from IT or Business will have likely gone through several online "Data Science boot camp" offerings. Roughly speaking, Data Science programs tend to emphasize the skills we summarized in Figure 19.1 under "Statistics" and "Programming". Business Analytics programs tend to de-emphasize these topics in favor of more instruction in aspects that would fall under "Business" in Figure 19.1.

What difference does your forecaster's background make in how they see the world?

- An IT person will be most familiar with the IT aspects of forecasting. They will think about data storage requirements, database setups, access latencies, runtimes, cloud vs. on-premise computations, data cleansing and preparation pipelines, etc.
- Someone who came to forecasting from the Business side of things will think in terms of what the data *means*. Is the behavior they observe in a time series normal or abnormal? Does it reflect new developments in the marketplace or issues with the supply chain? Can they deduce your salespeople's efforts from observed demands, and how can they take these into account when forecasting? "Business" forecasters are usually best able to understand and apply the differences between forecasting, targets, and plans.
- Data Scientists or Business Analysts usually have a solid grounding in the IT aspects of forecasting and almost as much understanding of the statistical foundations. Naturally, their IT knowledge will not be as deep as that of an IT veteran. Still, if you compare a Data Science or Business Analytics graduate with a Computer Science graduate, your Data Scientist/Business Analyst will usually be less deep but wider in their background. For a forecasting function, this is a sweet spot. Where Data Scientists are naturally usually weak – at least at first – is the business background. The *better* Data Scientists recognize this and are open to learning. Modern Business Analytics programs combine Data Science with a thorough understanding of business, making these candidates predestined for business forecasting functions.

Ideally, your forecasting team should contain diverse people with many different backgrounds, so they can learn from each other and combine their strengths. Forecasting is interdisciplinary; anyone with tunnel vision will be less effective.

Thinking about the background of your forecaster provides clues as to what may be vital for them. Few people *radically* change their outlook, starting as a marketing major and shifting to statistics and time series analysis in mid-career. Nevertheless, for someone to move into a forecasting position (and stay there) indicates several characteristics: they will be quantitatively oriented, be comfortable with ambiguity, and be at least somewhat interested in the business background.

19.3 Making your forecaster happy

On the one hand, the previous section should have given you a few ideas on what may be valuable to your forecaster. Make them happy by making sure they get the resources they need. Examples of "a forecaster's favorite resources" are:

- Up-to-date, clean, and usable data. Depending on your specific situation, investing some effort in cleaning up data or collecting (or buying!) additional data may yield good returns regarding improved forecast accuracy. Yes, this may require cooperation from IT functions outside the forecasting team. Getting this cooperation is one instance where a forecasting team's manager can add value.
- Computing resources. Chances are that your IT department is already budgeting for enough data storage space. Still, forecasting methods, especially the more modern ones, are data- and processing-cycle-hungry. Getting your forecaster enough computing resources may be a question of getting them a sufficiently powerful laptop – or an Amazon Web Services or Microsoft Azure subscription.
- Tools. Many forecasting tools, like R or Python, are free. But others aren't. Investing in a dedicated forecasting solution may be a worthwhile investment. Or not. Analyze the decision.
- Training. Forecasting, and Data Science in general, is a quickly evolving field. Without ongoing learning and training, the skillset of a forecaster will become obsolete quickly. Training may cost time and money. Staying current also requires the freedom to try out new ideas on existing data.
- Interaction. There are typically few forecasters in any given company, so your forecaster(s) may be interested in interacting with data scientists and forecasters from the outside. Such communications can infuse fresh thinking or give your forecaster a little recognition (which forecasters frequently do *not* get in-house, where good forecasts are rarely appreciated, and bad forecasts are remembered forever). Perhaps your forecaster would be interested in participating in Kaggle competitions, and you could support this with company time and computing resources. Or maybe they could disguise some of your data and organize a data science competition at Kaggle?

On the other hand, your forecaster may also need to consider development that is outside their comfort zone. They may be motivated to learn the newest Python module relevant to forecasting. But it may be much more helpful for them to attend a communications or presentation training or to shadow a business user for a week to understand what people actually *do* with their forecasts.

19.4 What can you expect from your forecaster?

If you take good care of your forecaster, you can expect them to take good care of your data. They will forecast what you need and explain their forecasts to the extent possible. (Machine learning methods are *hard* to explain, especially to a non-technical audience.) After all, forecasts that are not understood may not be trusted, and forecasts that are not trusted may not be used, so people may resort to making up their own forecasts. An organization that cannot agree on a common forecast is at a disadvantage; this is an outcome anyone near a dedicated forecasting function should avoid.

Your forecaster should also be able to dig into bad forecasts and at least make educated guesses as to why they were bad. Maybe the future differed sharply from the past. Or maybe there was a problem with the input data. If so, your forecaster should suggest possible improvements.

Your forecaster should also follow recent developments in the forecasting world. They do not need to participate in Kaggle competitions (although encourage them to do so!). Still, they should have an informed opinion of the methods used there and whether trying them on your data would be worthwhile. The answer to the question, "should we try a Gradient Boosting Machine to forecast our data?" may be, "our input data is so bad that applying cutting-edge tools to it will not be helpful – better to work on data quality first."

Your forecaster should provide meaningful input in any discussions with outside vendors of forecasting solutions. Of course, such a situation is always hard to navigate: is the forecaster afraid of becoming redundant or looking forward to getting new tools that will enable them to do their job more efficiently? Do they have the incentive to badmouth the external vendor or to argue for the solution with the bells and whistles that most caters to your forecaster's inner child? Nevertheless, your forecaster should be the person in the organization best qualified to understand and evaluate a vendor's claims, at least in forecasting performance, so listen closely to what your forecaster is telling you.

Conversely, here are two expectations you should *not* have of your forecaster:

- As we discuss elsewhere in this book, forecastability is always limited. Your forecaster cannot be expected to hit an arbitrary accuracy target. Telling your forecaster that "the industry standard" is a MAPE of 20%, so their bonus is contingent on them reaching 18% without regard to your data's specificities will only frustrate them (Kolassa 2008). The same holds if you take this year's accuracy and require next year's accuracy to improve on that. It may be possible – but posing this requirement *blindly* risks requiring your forecaster to forecast a fair coin with 60% accuracy.
- Similarly, there is frequently some wishful thinking that a new forecasting method – typically one involving Machine Learning – will yield significant advances in accuracy. Forecasters are often under pressure to wring precisely this accuracy improvement out of a new tool. Again, this may be feasible, or it may not be. You may very well have reached the limits of forecastability of your time series.

Key takeaways

1. Forecasting, like all data science, requires competency in four different domains:
   - Statistics
   - Programming/IT
   - Business
   - Communication
2. Forecasters who deeply understand all four dimensions are as rare as the proverbial unicorns. However, any well-rounded forecaster should be encouraged to develop in all four directions.
3. Your forecaster will likely have one of three different backgrounds:
   - IT
   - Business
   - Data Science/Business Analytics
4. Understanding your forecaster's background will help you understand through what kind of lens they perceive the world, what domains they may tend to favor in their development, and where learning outside their comfort zone may be helpful.
5. You can expect your forecasters to create sound forecasts, explain them to non-technical users, and help leverage them for maximum business benefit.
6. You should *not* expect your forecaster to hit arbitrary accuracy targets or to work wonders just because they got a new forecasting tool.

20

Sales and operations planning

We treat forecasting as a predominantly statistical exercise for most of this book. However, forecasting is as much an organizational exercise as a statistical one. Information is spread throughout the organization. The forecast has many stakeholders that require it as input to their planning processes. Thus, understanding demand forecasting requires understanding the statistical methods used to produce a forecast as well as how an organization creates a forecast and uses it for decision-making.

20.1 Forecasting organization

Managing this process is the realm of *Sales & Operations Planning (S&OP)*. Since S&OP has been around for a while, best practices for such processes are now established (e.g., Lapide 2014). There is an input and an output side to this process. On the input side, a good S&OP process supports sharing relevant information about the demand forecast. The emphasis here is on marketing and sales to share upcoming product launches, acquisition of new customers, planned promotions, and similar information with those responsible for forecasting. At the same time, other functions, such as operations, need to share relevant input data in the planning process, such as the inventory position, capacity available, etc. On the output side are coordinated plans based on the same input. Marketing develops a plan for promotions and demand management. Operations develops production and procurement plans. Finance develops cash flow plans and uses numbers coordinated with the other department to communicate with investors. Human resources develops a personnel plan based on the same forecast data.

Effective S&OP can coordinate the entire organization. Decision-makers will base their plans on all available information. If this process does not go well, functional areas hoard information, individual members influence forecasts, and the organization's functions lack coordination. The result will be highly inaccurate forecasts and, consequently, promotions without sufficient capacity and predictions shared with investors that bear little resemblance to actual plans, thus damaging the firm's credibility.

An S&OP process is a monthly process within an organization to enhance information sharing and coordinate plans. It usually involves a cross-functional team from marketing, operations, finance, sometimes human resources, and the dedicated forecasting team, if it exists. There are typically five different steps in an S&OP process. Beginning with *Data Gathering*, representatives from the various functions share relevant information and develop a standard set of business assumptions that goes into the forecast. In the actual *Demand Planning* stage, the team finalizes promotion and pricing decisions and agrees on a consensus forecast. Sales uses the consensus forecast to develop sales targets. Afterward, operations prepares inventory, production, capacity, and procurement decisions based on the consensus forecast during the *Supply Planning* stage.

Further, if the firm expects shortages, planners develop rationing and prioritization policies; they may also consider significant risk events, develop contingencies and share them with all team members. Finally, in the *Pre-Executive Meeting*, senior management can adjust any outcomes of the S&OP process. The finance function sometimes has veto power at this step to enable better cash flow planning and investor communications. Finally, top management discusses and finalizes all relevant plans during the *Executive Meeting*.

20.2 Organizational barriers

We must overcome two essential barriers to make this process work: incentives and organizational boundaries.

The first barrier stems from the fact that incentives across different functions are not aligned (see Section 16.3), and often no one is accountable for the quality of forecasts. Incentives for marketing or sales managers can lead them to lowball the forecast since they understand that their targets are often set depending on the forecast. Thus, lowering the forecast is an easy way for them to make their targets more obtainable. Or they may be incentivized to inflate the forecast since they understand this will push operations to create more inventory. More inventory decreases the chances of a stock-out happening and increases sales-related bonuses.

Operations managers are often compensated based on costs; one way to keep less inventory is to lowball the forecast and, thereby, lower production volumes. Finance will also chip in here since they will use the forecast to manage investor expectations. Decision-making and forecasting become mixed up. The forecast no longer coordinates activities but is a toy of organizational politics.

The second barrier is a result of different functional backgrounds and social identities. For example, people in marketing may have studied other topics than people in operations; they may have also had a different entry route into the organization. These experiences shape their way of perceiving and communicating organizational realities. Such unique thought worlds lead to challenges in managing a cross-functional team involving both groups.

Different groups may forecast in distinct units. Whereas finance forecasts in dollars of revenue, marketing may predict market shares, whereas operations is interested in product units. While we can convert these units, they represent a natural barrier to overcome. S&OP processes should standardize the conversion of these units.

Last but not least, having different functions always implies different social identities. Representatives from marketing will feel a natural allegiance to their function, as will the representatives from operations. While such social identification usually creates more trust within the group, it creates distrust between groups. Successfully managing a cross-functional S&OP team thus requires establishing standardized communication norms and breaking down functional barriers to build trust between the representatives of different functions.

One approach to overcome these barriers is to create a forecasting group that is organizationally separate from all other participating functions; this promotes accountability for forecast accuracy and a social identity focused on forecasting. It also creates professionalism and allows compensating people based on forecasting performance. A survey on forecasting practice found that 38% of responding organizations have introduced a separate group for forecasting, and 62% of those forecasting groups owned the forecasting process (McCarthy et al. 2006). We can compensate forecasters in such a group based on the accuracy of forecasts without creating asymmetry in their incentive systems; incentivizing forecasting in all other areas requires meticulous calibration to offset the existing incentives in these areas to over- or under-forecast (Scheele, Thonemann, and Slikker 2017). If creating a separate forecasting function is impossible, the people participating in S&OP planning should de-emphasize their functional association. For example, it may be possible to remove the participating employees from their functional incentive systems and reward them according to firm performance or forecasting performance instead.

Another essential aspect of rational S&OP is to demystify the forecasts; all those involved in the forecasting process should be familiar with the data, software, and algorithms used in creating forecasts. Assumptions should be documented and transparent to everyone involved, and individuals should be held accountable for their judgment. As we emphasize in Chapter 16, in the long run, it is always possible to tell whether adjustments made to a statistical forecast were helpful or hurtful for forecasting performance through Forecast Value Added analysis.

An exciting avenue to resolve incentive issues in S&OP is to use past forecast accuracy of forecasts from different functional areas to determine the future weight given to these different functional forecasts when calculating a consensus. Under such a system, functions that bias their forecast will quickly lose their ability to influence the consensus forecast, incentivizing them to avoid biasing it.

A good case study of transforming an S&OP process is given by Oliva and Watson (2009). The authors study an electronics manufacturer that started with a dysfunctional forecasting process; the company had three different functions (Sales, Operations, and Finance), creating three different forecasts; the only information sharing happened in non-standardized spreadsheets and hallway conversations. The company proceeded to generate a process that started by (1) creating a separate group that was responsible for the statistical side of forecasting, (2) creating a common assumptions package where each function would contribute crucial information about the development of the business, (3) allowing the different functions to generate separate forecasts based on the same information, but then (4) integrating these forecasts using a weighted average, where the weight attributed to each function depended on past accuracy, and (5) limiting revisions to this initial consensus forecast to only those instances where actual data could be brought up to support any modifications to the forecast. The result was a stark increase in forecasting performance. Whereas company forecasts had an accuracy (1-MAPE) of only 50% before this re-design, this accuracy jumped to almost 90% after re-structuring the S&OP process.

Key takeaways

1. Forecasting is as much a social and an organizational activity as a statistical one.
2. Employees' functional backgrounds may incentivize them to influence the forecast in ways that are not optimal for the business. Consider changing their incentives to align with getting an unbiased forecast.
3. It may be very beneficial to separate forecasting out from traditional functional areas organizationally.
4. Similarly, employees' backgrounds may influence how they think and communicate about forecasts. Keep this in mind and work toward open communication.

21

Why does forecasting fail?

It is easy to become frustrated with forecasting. Sometimes the forecasting process seems to "fail." Let us discuss what failure means in the context of forecasting. Many topics we picked here have already been discussed elsewhere in the book, but it is still useful to bring them together in one place.

21.1 There is no alternative to forecasting

When we say something "fails," the question is, "compared to what?" You can only fail an exam because there is a reasonable possibility of succeeding at it. We do not say that unaided humans "fail at flying" simply because there is no expectation that humans can fly without equipment.

In this sense, forecasting does not fail, because we have no alternative. We need to plan for the future; to successfully plan, we need expectations. These expectations are forecasts, whether we call them that or not. Our expectations may have a lot of uncertainty. Maybe we believe our product could sell like crazy or not at all. Still, we must use *some* expectation framework to plan. Even a make-to-order production process relies on implicit forecasts, namely that customer demand and our supply chain will stay sufficiently stable for this simple process to continue operating satisfactorily. If a sudden significant spike in demand is imminent, we may change to make-to-stock, at least for part of our assortment. A crucial part of making such decisions is to form proper expectations, i.e., forecasting.

Thus, the question is not whether we should forecast or not. We have no choice. It is much more constructive to ask whether we can efficiently improve our forecasts.

21.2 What we can forecast

In principle, we can forecast anything, but how accurately? We can predict some demand patterns with a high degree of accuracy. Others elude us. Why are some demand patterns more challenging to forecast accurately than others?

A critical step in forecasting is understanding when we can forecast something accurately and when forecasts might not be better than blind chance. The forecastability (see Section 5.5) of any demand depends on several factors, including (Hyndman 2021):

- How well we understand the factors that contribute to its variation
- How much data is available
- Whether the forecast can affect itself via feedback
- Whether the future is similar to the past
- How much natural or inexplicable variation there is
- How far into the future do we forecast

For instance, when forecasting the number of admissions to Accident and Emergency (A&E) Departments in the next two days, most factors on this list work in our favor. We know the key drivers of admissions demand. For example, admissions demand exhibits both hour-of-the-day and day-of-the-year seasonality. Hospitals generally have access to a long history of data on admissions. With the right skills, we can develop a good forecasting model linking admissions demand to key demand drivers. Our forecasts can be accurate. However, if a manager in the A&E department requires hourly forecasts for a longer horizon, producing such forecasts becomes more challenging.

In late 2019, the COVID-19 pandemic hit with several devastating effects on hospital service providers, thereby changing admission patterns. The future was no longer similar to the past in A&E departments. Consequently, historical data contained less information about the future. Forecast accuracy decreased as a result.

Forecasting a currency exchange rate is an arduous task. While lots of data is available on past currency rates, we do not understand the causal drivers of variation. Most importantly, forecasts of the future exchange rate influence the exchange rate itself via feedback: if we forecast the price of a currency to rise in the future, we buy it today and thus bid up the price. Forecasts become somewhat self-fulfilling prophecies. The forecast will affect people's behaviors. In such situations, forecasters must be keenly aware of their limitations and impact.

The best we can do in forecasting is to capture the systematic structure or causal drivers contributing to the variation of the demand, find a forecasting model that accurately represents that structure, and then hope that demand characteristics do not change in the future. Understanding all factors contributing to the variation of demand is fundamental to producing an accurate demand forecast. Domain or business knowledge can play a crucial role.

21.3 We can't achieve unlimited accuracy

Our forecasting accuracy is always limited (see also Section 5.5). If we want to forecast a fair coin toss, there is simply no way to get better accuracy than 50%, and the same holds in forecasting any time series. Thus, if we require a forecast to be better than what is possible, we have a problem. Every casino visitor faces this problem. We can't forecast the number in American roulette with a success probability better than 1/38. Still, the casino pays out at a rate of 36:1 only, so our forecasts are not good enough to make money reliably.

If we know the momentum and spin at which the roulette ball was tossed, shouldn't we be able to predict its result more accurately? Yes, and for high-stakes forecasts, collecting more and more information exactly this way makes sense. The challenge lies in determining whether one has reached the end of reasonable forecast accuracy improvements (Tim 2017) – in other words, whether the resulting forecast accuracy improvement is worth collecting more data, which is usually expensive. In addition, we need to keep in mind that more complex models may even lead to worse forecasts (see Section 11.7). Also, don't be surprised if the casino asks you to leave if you set up high-end physical measurement equipment around their roulette wheel. Card counting, an easy way to increase the accuracy of your forecasts in blackjack, is considered illegal in casinos, after all.

21.4 Confusing forecasts, targets, decisions and plans

You may sometimes hear statements like *we need to hit the forecast*, which suggests a misunderstanding between a goal and a forecast. A forecast is not the same as a goal. A forecast is an honest assessment of future demand based on all the past and future information available when generating the forecast. A goal is an outcome that we strive to accomplish. We use goals to motivate and coordinate people.

Forecasting should be an integral part of any decision-making process, whether on an operational, tactical, or strategic level. It is easy to confuse a decision variable with a forecast, but rather, forecasts inform decisions. For instance, a government body will use the electricity demand forecast in the next 20 years to decide whether to build a new power plant. A healthcare provider may forecast the doses administrated for next month to be 60,000, but they may decide to keep 70,000 doses in the cold chain warehouse to avoid a missed opportunity. "70,000 doses" is not a forecast. It is a decision.

Plans are responses to forecasts, decisions, and goals. Planning involves determining the appropriate actions required to achieve targets, as informed by forecasts, following the decision-making process.

21.5 No systematic tracking of forecast quality

Decision-makers often claim that forecasts are not good enough. However, frequently nobody knows just *how* good they are because no one tracks their accuracy systematically. A forecast is rarely spot on, and we should not expect it to be. The first step in assessing a forecast's quality is using a well-established error measure. We can track this measure over time to evaluate whether accuracy deteriorates or improves or whether some forecasts are systematically more accurate than others.

21.6 Inappropriate error measures

We should tailor our error measure to the decision the forecast is supposed to support. For example, if we use our forecast to drive replenishment, we would calculate quantile forecasts and assess these using a pinball loss. Of course, we should not evaluate a quantile loss using the MAPE. And as a matter of fact, it may be yet better to not assess forecast accuracy at all, but the resulting business outcome in terms of overstock and stockout rates, and try to get a handle on how problematic forecasts contribute to any problem here – or whether they even do so at all (Kolassa 2023b). Thus, the forecast and its evaluation measures must all be tailored to the consuming process.

21.7 Data availability and quality

We can often improve forecasts by understanding the underlying drivers and leveraging better data. If your product is a commodity and price strongly drives demand, then models that do not include the price as a predictor will not be accurate. If you forecast demand for IT support call centers, releasing a new version of the product you are supporting will generate many calls. Include release dates in your forecast.

Data may be *available*, but its quality may not be sufficient (compare Section 5.3). And it may be expensive, illegal, or even impossible to obtain better quality data on predictors that would help us improve forecasts. Knowing our customers' future plans might help us plan better, and such information sharing does happen. But we should expect our customers to ask what they get in return for providing us with this data. Information sharing will cost us beyond the technical effort to set up the data feed. In contrast, knowing our competitors' plans might help us forecast better, but obtaining non-public information might well be illegal. Finally, knowing exactly when the first sunny weekend in spring (or the first snowstorm in winter) happens would help us enormously with stocking related products – but weather forecasts are not accurate far out enough to be helpful.

21.8 Too much judgmental intervention

Humans adjust most statistical forecasts before their use for subsequent decisions, e.g., in alignment meetings in an S&OP process. Sometimes such judgmental interventions add value, e.g., when the human knows of a driver that the forecasting system does not use. However, humans are notoriously good at seeing patterns where none exist and are prone to change the forecast to fit perceived patterns. They add noise and make the forecast worse in the process (see Sections 16.1 and 16.3).

Even judgmental adjustments made for good reasons can make things worse. For instance, when a big promotion is coming up, it may make sense to adjust the forecast upward. However, if the system has already accounted for the promotion, our adjustment may overshoot it. The same may happen if someone else has already adjusted the forecast for this promotion before us or if someone else does so again after us.

It is a good idea to test whether judgmental interventions improve forecasts by running a Forecast Value Added analysis (see Section 5.5). You can use this process to assess the value of any forecast improvement effort, e.g., by evaluating the error of one model without a predictor and a more complex model that includes the predictor, or by comparing the forecasts before and after they have been judgmentally adjusted.

21.9 Follow-on processes do not leverage forecasts

Decision-making processes can be disconnected from the forecast that should influence them. On the one hand, this may be because forecasts are judgmentally adjusted to the point when they are devoid of reality (see the section above), and are therefore ignored by the people who should in principle rely on them.

Sometimes, however, factors besides the forecast dominate the decisions. For instance, we may be ordering based on our demand forecast – but the minimum order amount is so high that we essentially always place very high orders when current stocks fall below some level. In other words, the safety stock is not determined by forecast uncertainty, but by minimum order agreed on between our suppliers and our purchasing department. In such a situation, forecast accuracy may not have any impact (within bounds). Investing a lot of resources into improving forecasts would make little sense. Instead, we should either try to change the logistical framework to be more adaptive so that our forecasts make a difference, or concentrate on other products where the forecasts matter more.

21.10 Feedback to the process being forecasted

Some forecasts feed back into the decisions that use them. For example, suppose we forecast the demand for intensive care beds during a pandemic. If our forecast shows that demand will outstrip supply, that may lead governments to impose strict lockdowns, which would, in turn, break the spread of the pandemic and reduce the need for hospital beds we are forecasting (Goodwin 2023). Similarly, our forecasts may indicate that demand for a particular product is too low to use up our stock before the product becomes obsolete. In response, we might reduce prices or run promotions to stimulate demand, which will then hopefully come in higher than originally forecasted. This is of course not a failure of the forecast!

Key takeaways

1. There is no alternative to forecasting. The critical question is whether we do a good job at forecasting.
2. Whether forecasts are good or bad depends on the decisions we use a forecast for and the forecastability of the time series.

3. Forecasts can only be as good as the predictor data we feed them. There is a point of diminishing marginal returns in improving input data and models.
4. Forecast Value Added analysis can help us find value-adding and value-destroying steps in our process.
5. Assessing accuracy can be very challenging if the forecast feeds back into decisions that influence demand.

Learning more

22

Resources

This chapter summarizes additional resources for you to learn more about forecasting. We categorize these resources by grouping them into different sections for non-technical, somewhat technical, and technical material. The present book would be in the "somewhat technical" section. The resources in the "technical" section require more technical background knowledge, typically in terms of statistics. We also point to forecasting organizations, events, datasets, and online resources.

22.1 Non-technical material

Gilliland (2010) focuses on the process aspects of forecasting. In our opinion, its most valuable parts are two chapters on worst practices in forecasting and one chapter on Forecast Value Added analysis, a framework that aims to capture how much value a forecast (or specific steps in a forecasting process) adds in a business sense (see Section 16.3).

Goodwin (2017) summarizes what Professor Paul Goodwin at the University of Bath has learned about forecasting in his decades-long career studying the topic. It paints a broad canvas and provides insights into many different areas of forecasting.

Tetlock and Gardner (2015) describe a sociological experiment on forecasting. Are there people who can forecast very well? If so, how can we help them become even better? The intellectual origins of this book lie in the "Wisdom of Crowds" phenomenon, i.e., the argument that forecasts should rely on average predictions from a group rather than the opinions of individual experts. Somewhat ironically, this book concludes that so-called "super forecasters" exist, i.e., people who are remarkably accurate in their predictions. The book focuses less on demand than geopolitical forecasting (the CIA enabled this experiment). While there is no immediate applicability to better demand forecasting, the book is insightful and a joy to read.

Vandeput (2023) makes very similar points to us in his collection of best practices in demand forecasting. His description of what to look out for and what pitfalls to avoid is informed by many years of practice and experience "in the trenches" of supply chain forecasting. This book makes for an excellent companion to the one you have just read.

Foresight: The International Journal of Applied Forecasting (https://forecasters.org/foresight/) is a quarterly publication of the International Institute of Forecasters aimed at practitioners. Articles are written for and by practitioners. Academics occasionally contribute, but they write their articles with a practitioner audience in mind. The journal frequently publishes features, i.e., a longer article on a specific topic and commentaries by other authors. It also publishes books compiling articles on a common theme, e.g., Gilliland, Tashman, and Sglavo (2015) and Gilliland, Tashman, and Sglavo (2021). (Full disclosure: one of the present authors, SK, is a Deputy Editor at *Foresight*.)

22.2 Somewhat technical material

Kolassa and Siemsen (2016) is an earlier book by two of the authors of the book you are currently reading. We based our present work on it but expanded it significantly.

Hyndman and Athanasopoulos (2021) provide a more technical introduction to forecasting. The authors are two of the foremost academic forecasting experts of the day. If you want to know how to derive ARIMA orders from an ACF/PACF plot or how optimal hierarchical reconciliation of forecasts works, then this book is for you. While it is more technical, understanding the content of this book only requires (some) matrix algebra. Readers do not need to have a deep knowledge of statistics. The book is also freely available online in two different editions, which are based on two flavors of the free statistical computing platform R (R Core Team 2022): the 2nd edition (https://otexts.com/fpp2/) uses base R and the `forecast` package (Hyndman et al. 2023), whereas the 3rd edition (https://otexts.com/fpp3/) uses the tidyverse (with a somewhat steeper learning curve than base R) and the newer `fable` package (O'Hara-Wild et al. 2020).

Shmueli and Lichtendahl (2018) is an introductory textbook that also focuses on forecasting with R, with many worked examples and exercises. There is also an analogous textbook (Shmueli 2016), which instead uses Microsoft Excel and the XLMiner add-on.

Ord, Fildes, and Kourentzes (2017) is a software-agnostic textbook that is a little more verbose and business-user-friendly than Hyndman and Athanasopoulos (2021).

22.3 Technical material

If you want to get up to speed with the latest developments in forecasting, your best bet is to look at the *International Journal of Forecasting* (IJF; https://www.sciencedirect.com/journal/international-journal-of-forecasting), published by the International Institute of Forecasters (see below). Most articles are abstract but at the cutting edge of research in forecasting. The journal publishes regular special issues dedicated, for example, to the M forecasting competitions. (The similarly named *Journal of Forecasting*, https://onlinelibrary.wiley.com/journal/1099131x, focuses much more on forecasting other processes than demand, e.g., financial or macroeconomic time series.)

One recent paper published in the *IJF* is the comprehensive review article by Petropoulos, Apiletti et al. (2022). Almost 80 worldwide experts pooled their knowledge to describe the current state of the art in forecasting, for demand as well as for other applications. The review includes ample references to recent academic papers. The article is published online at https://forecasting-encyclopedia.com/, and the authors plan on keeping this online version current.

Experts in Machine Learning applying their skills to forecasting are often unaware of the specific challenges in this sub-field. Hewamalage, Ackermann, and Bergmeir (2022) provide a valuable introduction to the pitfalls in forecast evaluation, geared to this group.

22.4 Non-profit organizations

Various non-profit organizations have a mission to expand knowledge about forecasting.

The International Institute of Forecasters (IIF; https://forecasters.org/) is a global association of academics, practitioners, and consultants in forecasting. It brings together all sides involved in forecasting, allowing mutual learning. The IIF publishes two journals, the more practitioner-oriented *Foresight: The International Journal of Applied Forecasting* and the more academic *International Journal of Forecasting* (see above for both), and organizes an annual conference, the International Symposium on Forecasting (ISF; see below). The IIF also certifies courses in forecasting at various universities around the world.

The Centre for Marketing Analytics and Forecasting (CMAF; https://www.lancaster.ac.uk/lums/research/areas-of-expertise/centre-for-marketing-analytics-and-forecasting/) at Lancaster University Management School in the UK is the most important center of excellence in forecasting worldwide. It is an academic organization, but it offers consulting services and educates professionals in forecasting. It also provides very well-trained students for forecasting projects.

22.5 Events

The most relevant event in forecasting is the annual International Symposium on Forecasting (ISF; https://isf.forecasters.org/). This global conference is organized by the International Institute of Forecasters (IIF; see above). Attendees are an eclectic mix of academics and practitioners. There are dedicated practitioner tracks, but the academic tracks are also worth attending. Various workshops are offered before the conference as well.

22.6 Datasets

Kaggle (https://www.kaggle.com/) is a general data science platform that regularly organizes competitions, some of which are forecasting-themed. You can browse through the forums to find out how people address an ongoing or finished challenge, download data to try your algorithms, or even contact Kaggle to have them organize a forecasting competition using your data!

The Monash Time Series Forecasting Repository at https://forecastingdata.org/ (Godahewa et al. 2021) contains datasets of *related* time series, which are especially suitable for testing hierarchical forecasting techniques as in Chapter 13.

Many packages for the free statistical computing platform R (R Core Team 2022) contain time series datasets. For instance, the `fpp2` (Hyndman 2020) and the `fpp3` packages (Hyndman 2023) contain the datasets used in the 2nd and 3rd edition of Hyndman and Athanasopoulos (2021) (see above), and more data can be found in the `forecast` (Hyndman et al. 2023), the `fable` (O'Hara-Wild et al. 2020) and the `tsdl` (Hyndman and Yang 2023) packages. The `Mcomp` package (Hyndman 2018) contains the datasets used in the M1 and the M3 competition. The data for the M5 competition can be found at https://www.kaggle.com/c/m5-forecasting-accuracy. (The recently concluded M6 competition used financial time series and is thus a little less helpful for *demand* forecasting.)

Many more datasets can be obtained from national statistical offices, or by a simple internet search. For instance, you can download time series of posts to the StackExchange network through the StackExchange Data Explorer (https://data.stackexchange.com/). Finally, if you need a very specific dataset, you can always ask at the OpenData StackExchange site (https://opendata.stackexchange.com/).

22.7 Online resources

Of course, most resources described above have an online presence, and we noted the URL wherever applicable. There are also a few purely online sites of interest to forecasters.

The CMAF mentioned above organizes a regular schedule of forecasting-themed live-streamed webinars, the Friday Forecasting Talks (FFTs; https://cmaf-fft.lp151.com/). Past webinars are archived on the CMAF YouTube channel (https://www.youtube.com/@lancastercmaf).

The IIF (see above) publishes a series of podcasts called "Forecasting Impact", which features interviews with forecasting experts. Past episodes are archived at https://forecasters.org/publications/forecasting-impact-podcast/.

Finally, an invaluable resource for the practicing forecaster is CrossValidated (https://stats.stackexchange.com), a Q&A site for statistics with many forecasting questions (and answers). With over 200,000 questions, your specific forecasting question may already have a response, or you could start a new thread (just *please* read the site help first: https://stats.stackexchange.com/help).

Key takeaways

1. There are lots of resources to improve your forecasting – but don't forget the other facets that make good forecasters (see Section 19.1).
2. Happy forecasting!

22.7 Online resources

References

Ali, Mohammad M., John E. Boylan, and Aris A. Syntetos. 2012. "Forecast Errors and Inventory Performance Under Forecast Information Sharing." *International Journal of Forecasting* 28 (4): 830–41. https://doi.org/10.1016/j.ijforecast.2010.08.003.

Anderson, Eric T., Gavan J. Fitzsimons, and Duncan Simester. 2006. "Measuring and Mitigating the Costs of Stockouts." *Management Science* 52 (11): 1751–63. https://doi.org/10.1287/mnsc.1060.0577.

Andy T. 2016. "Is it unusual for the MEAN to outperform ARIMA?" Cross Validated. https://stats.stackexchange.com/q/124955.

Angrist, Joshua D., and Jörn-Steffen Pischke. 2009. *Mostly Harmless Econometrics*. Princeton University Press. https://doi.org/10.1515/9781400829828.

Armstrong, J. Scott. 2001. "Combining Forecasts." In *Principles of Forecasting – A Handbook for Researchers and Practitioners*, edited by J. Scott Armstrong, 1–19. Kluwer.

Arnold, Jeffrey B. 2021. *Ggthemes: Extra Themes, Scales and Geoms for 'Ggplot2'. R Package Version 4.2.4*. https://CRAN.R-project.org/package=ggthemes.

Athanasopoulos, George, Rob J. Hyndman, Nikolaos Kourentzes, and Fotios Petropoulos. 2017. "Forecasting with Temporal Hierarchies." *European Journal of Operational Research* 262 (1): 60–74. https://doi.org/10.1016/j.ejor.2017.02.046.

Athanasopoulos, George, Rob J. Hyndman, Haiyan Song, and Doris C. Wu. 2011. "The Tourism Forecasting Competition." *International Journal of Forecasting* 27 (3): 822–44. https://doi.org/10.1016/j.ijforecast.2010.04.009.

Bandara, Kasun, Rob J. Hyndman, and Christoph Bergmeir. 2021. "MSTL: A Seasonal-Trend Decomposition Algorithm for Time Series with Multiple Seasonal Patterns," July. https://arxiv.org/abs/2107.13462.

Batchelor, Roy. 2010. "Worst-Case Scenarios in Forecasting: How Bad Can Things Get?" *Foresight: The International Journal of Applied Forecasting* 18: 27–32.

Benesty, Jacob, Jingdong Chen, Yiteng Huang, and Israel Cohen. 2009. "Pearson Correlation Coefficient," 1–4. https://doi.org/10.1007/978-3-642-00296-0_5.

Berry, Tim. 2010. *Sales and Market Forecasting for Entrepreneurs*. Business Expert Press.

Biais, Bruno, and Martin Weber. 2009. "Hindsight Bias, Risk Perception, and Investment Performance." *Management Science* 55 (6): 1018–29. https://doi.org/10.1287/mnsc.1090.1000.

Boulaksil, Youssef, and Philip Hans Franses. 2009. "Experts' Stated Behavior." *Interfaces* 39 (2): 168–71. https://doi.org/10.1287/inte.1080.0421.

Boylan, John E., and M. Zied Babai. 2016. "On the Performance of Overlapping and Non-Overlapping Temporal Demand Aggregation Approaches." *International Journal of Production Economics* 181, Part A: 136–44. https://doi.org/10.1016/j.ijpe.2016.04.003.

Boylan, John E., and Aris A. Syntetos. 2006. "Accuracy and Accuracy-Implication Metrics for Intermittent Demand." *Foresight: The International Journal of Applied Forecasting* 4: 39–42.

———. 2021. *Intermittent Demand Forecasting: Context, Methods and Applications.* Wiley. https://www.ebook.de/de/product/23546327/john_e_boylan_aris_a_syntetos_intermittent_demand_forecasting_context_methods_and_applications.html.

Brau, Rebekah, John Aloysius, and Enno Siemsen. 2023. "Demand Planning for the Digital Supply Chain: How to Integrate Human Judgment and Predictive Analytics." *Journal of Operations Management.* https://doi.org/10.1002/joom.1257.

Chatfield, Chris. 2001. "Prediction Intervals for Time-Series Forecasting." In *Principles of Forecasting – A Handbook for Researchers and Practitioners*, edited by J. Scott Armstrong, 475–94. Kluwer.

———. 2007. "Confessions of a Pragmatic Forecaster." *Foresight: The International Journal of Applied Forecasting* 6: 3–9.

Chatfield, Chris, Anne B. Koehler, J. Keith Ord, and Ralph D. Snyder. 2001. "A New Look at Models for Exponential Smoothing." *Journal of the Royal Statistical Society: Series D (The Statistican)* 50 (2): 147–59. https://doi.org/10.1111/1467-9884.00267.

Choi, Hyunyoung, and Hal Varian. 2012. "Predicting the Present with Google Trends." *Economic Record* 88 (s1): 2–9. https://doi.org/10.1111/j.1475-4932.2012.00809.x.

Claeskens, Gerda, Jan R. Magnus, Andrey L. Vasnev, and Wendun Wang. 2016. "The Forecast Combination Puzzle: A Simple Theoretical Explanation." *International Journal of Forecasting* 32 (3): 754–62. https://doi.org/10.1016/j.ijforecast.2015.12.005.

Clarke, Simon. 2006. "Managing the Introduction of a Structured Forecast Process: Transformation Lessons from Coca-Cola Enterprises Inc." *Foresight: The International Journal of Applied Forecasting* 4: 21–25.

Clemen, Robert T. 1989. "Combining Forecasts: A Review and Annotated Bibliography." *International Journal of Forecasting* 5 (4): 559–83. https://doi.org/10.1016/0169-2070(89)90012-5.

Cleveland, Robert B, William S Cleveland, Jean E McRae, and Irma Terpenning. 1990. "STL: A Seasonal-Trend Decomposition." *Journal of Official Statistics* 6 (1): 3–73.

Corsten, Daniel, and Thomas Gruen. 2004. "Stock-Outs Cause Walkouts." *Harvard Business Review*, May, 26–28.

Croston, J D. 1972. "Forecasting and Stock Control for Intermittent Demands." *Operational Research Quarterly* 23 (3): 289–303. https://doi.org/10.1057/jors.1972.50.

D'Aveni, Richard. 2015. "The 3-d Printing Revolution." *Harvard Business Review* 93 (5): 40–48.

De Livera, Alysha M., Rob J. Hyndman, and Ralph D. Snyder. 2011. "Forecasting Time Series with Complex Seasonal Patterns Using Exponential Smoothing." *Journal of the American Statistical Association* 106 (496): 1513–27. https://doi.org/10.1198/jasa.2011.tm09771.

Dietvorst, Berkeley J., Joseph P. Simmons, and Cade Massey. 2015. "Algorithm Aversion: People Erroneously Avoid Algorithms After Seeing Them Err." *Journal of Experimental Psychology: General* 144 (1): 114–26. https://doi.org/10.1037/xge0000033.

Engle, Robert F. 2001. "GARCH 101 : The Use of ARCH/GARCH Models in Applied Econometrics." *Journal of Economic Perspectives* 15 (4): 157–68. https://doi.org/10.1257/jep.15.4.157.

Facebook's Core Data Science Team. 2022. *Prophet: Forecasting at Scale. Python package version 1.1.1.* https://facebook.github.io/prophet/.

Fahimnia, Behnam, Meysam Arvan, Tarkan Tan, and Enno Siemsen. in press. "A Hidden Anchor: The Influence of Service Levels on Demand Forecasts." *Journal of Operations Management*, 2022. https://doi.org/10.1002/joom.1229.

Federico Garza, Cristian Challú, Max Mergenthaler Canseco. 2022. "StatsForecast: Lightning Fast Forecasting with Statistical and Econometric Models." PyCon Salt Lake City, Utah, US 2022. https://github.com/Nixtla/statsforecast.

Fildes, Robert, Paul Goodwin, Michael Lawrence, and Konstantinos Nikolopoulos. 2009. "Effective Forecasting and Judgmental Adjustments: An Empirical Evaluation and Strategies for Improvement in Supply-Chain Planning." *International Journal of Forecasting* 25 (1): 3–23. https://doi.org/10.1016/j.ijforecast.2008.11.010.

Fildes, Robert, Stephan Kolassa, and Shaohui Ma. 2022. "Post-script – Retail forecasting: Research and practice." *International Journal of Forecasting* 38 (4): 1319–24. https://doi.org/10.1016/j.ijforecast.2021.09.012.

Fildes, Robert, Shaohui Ma, and Stephan Kolassa. 2022. "Retail Forecasting: Research and Practice." *International Journal of Forecasting* 38 (4): 1283–1318. https://doi.org/10.1016/j.ijforecast.2019.06.004.

Fildes, Robert, and Fotios Petropoulos. 2015. "Improving Forecast Quality in Practice." *Foresight: The International Journal of Applied Forecasting* 36: 5–12.

Fisher, Aaron, Cynthia Rudin, and Francesca Dominici. 2019. "All Models Are Wrong, but *Many* Are Useful: Learning a Variable's Importance by

Studying an Entire Class of Prediction Models Simultaneously." *Journal of Machine Learning Research* 20: 1–81.

Fotios Petropoulos, Yael Grushka-Cockayne, Enno Siemsen, and Evangelos Spiliotis. 2023. "Wielding Occam's Razor: Fast and Frugal Retail Forecasting." *Working Paper.*

Franses, Philip Hans. 2016. "A Note on the Mean Absolute Scaled Error." *International Journal of Forecasting* 32 (1): 20–22. https://doi.org/10.1016/j.ijforecast.2015.03.008.

Friedman, Jerome H. 2001. "Greedy Function Approximation: A Gradient Boosting Machine." *The Annals of Statistics* 29 (5): 1189–1232. https://doi.org/10.1214/aos/1013203451.

———. 2002. "Stochastic Gradient Boosting." *Computational Statistics & Data Analysis* 38 (4): 367–78. https://doi.org/10.1016/s0167-9473(01)00065-2.

Gardner, Everette Shaw, Jr. 2006. "Exponential Smoothing: The State of the Art – Part II." *International Journal of Forecasting* 22 (4): 637–66. https://doi.org/10.1016/j.ijforecast.2006.03.005.

Gardner, Everette Shaw, Jr., and Ed. Mckenzie. 1985. "Forecasting Trends in Time Series." *Management Science* 31 (10): 1237–46. https://doi.org/10.1287/mnsc.31.10.1237.

Gelper, Sarah, Roland Fried, and Christophe Croux. 2009. "Robust Forecasting with Exponential and Holt-Winters Smoothing." *Journal of Forecasting*, 285–300. https://doi.org/10.1002/for.1125.

Géron, Aurélien. 2019. *Hands-on Machine Learning with Scikit-Learn, Keras, and TensorFlow.* 2nd ed. O'Reilly UK Ltd.

Gilliland, Michael. 2010. *The Business Forecasting Deal.* Hoboken, NJ: John Wiley & Sons.

———. 2013. "FVA: A Reality Check on Forecasting Practices." *Foresight: The International Journal of Applied Forecasting* 29: 14–18.

Gilliland, Michael, Len Tashman, and Udo Sglavo, eds. 2015. *Business Forecasting: Practical Problems and Solutions.* Wiley. https://www.ebook.de/de/product/25014190/michael_gilliland_len_tashman_udo_sglavo_business_forecasting.html.

———, eds. 2021. *Business Forecasting: The Emerging Role of Artificial Intelligence and Machine Learning.* Hoboken, NJ: Wiley.

Godahewa, Rakshitha, Christoph Bergmeir, Geoffrey I. Webb, Rob J. Hyndman, and Pablo Montero-Manso. 2021. "Monash Time Series Forecasting Archive." In *Neural Information Processing Systems Track on Datasets and Benchmarks.*

Goodwin, Paul. 2000. "Correct or Combine? Mechanically Integrating Judgmental Forecasts with Statistical Methods." *International Journal of Forecasting* 16 (2): 261–75. https://doi.org/10.1016/s0169-2070(00)00038-8.

———. 2017. *Forewarned – a Sceptic's Guide to Prediction.* Biteback Publishing.

———. 2023. "Should We Always Use Forecasts When Facing the Future?" *Foresight: The International Journal of Applied Forecasting* 69: 20–22.

Goodwin, Paul, and Richard Lawton. 1999. "On the Asymmetry of the Symmetric MAPE." *International Journal of Forecasting* 15 (4): 405–8. https://doi.org/10.1016/S0169-2070(99)00007-2.

Gould, Phillip G., Anne B. Koehler, J. Keith Ord, Ralph D. Snyder, Rob J. Hyndman, and Farshid Vahid-Araghi. 2008. "Forecasting Time Series with Multiple Seasonal Patterns." *European Journal of Operational Research* 191 (1): 207–22. https://doi.org/10.1016/j.ejor.2007.08.024.

Green, Kesten, and Len Tashman. 2008. "Should We Define Forecast Error as $e = F - A$ or $e = A - F$?" *Foresight: The International Journal of Applied Forecasting* 10: 38–40.

———. 2009. "Percentage Error: What Denominator." *Foresight: The International Journal of Applied Forecasting* 12: 36–40.

Hanley, James A., Lawrence Joseph, Robert W. Platt, Moo K. Chung, and Patrick Belisle. 2001. "Visualizing the Median as the Minimum-Deviation Location." *The American Statistician* 55 (2): 150–52. https://doi.org/10.1198/000313001750358482.

Haran, Uriel, Don A. Moore, and Carey K. Morewedge. 2010. "A Simple Remedy for Overprecision in Judgment." *Judgment and Decision Making* 5 (7): 467–76. https://doi.org/10.1017/S1930297500001637.

Harrell, Frank E., Jr. 2015. *Regression Modeling Strategies: With Applications to Linear Models, Logistic and Ordinal Regression, and Survival Analysis.* 2nd ed. Springer Series in Statistics. Springer. https://doi.org/10.1007/978-3-319-19425-7.

Harvey, Nigel. 1995. "Why Are Judgments Less Consistent in Less Predictable Task Situations?" *Organizational Behavior and Human Decision Processes* 63 (3): 247–63. https://doi.org/10.1006/obhd.1995.1077.

Harvey, Nigel, Teresa Ewart, and Robert West. 1997. "Effects of Data Noise on Statistical Judgement." *Thinking & Reasoning* 3 (2): 111–32. https://doi.org/10.1080/135467897394383.

Hastie, Trevor J, and Robert J Tibshirani. 1990. *Generalized Additive Models.* Vol. 43. CRC Press.

Hewamalage, Hansika, Klaus Ackermann, and Christoph Bergmeir. 2022. "Forecast Evaluation for Data Scientists: Common Pitfalls and Best Practices." *Data Mining and Knowledge Discovery*, December. https://doi.org/10.1007/s10618-022-00894-5.

Hill, Arthur V., Weiyong Zhang, and Gerald F. Burch. 2015. "Forecasting the Forecastability Quotient for Inventory Management." *International Journal of Forecasting* 31 (3): 651–63. https://doi.org/10.1016/j.ijforecast.2014.10.006.

Hoover, Jim. 2006. "Measuring Forecast Accuracy: Omissions in Today's Forecasting Engines and Demand-Planning Software." *Foresight: The International Journal of Applied Forecasting* 4: 32–35. https://ideas.repec.org/a/for/ijafaa/y2006i4p32-35.html.

Hyndman, Rob J. 2006. "Another Look at Forecast-Accuracy Metrics for Intermittent Demand." *Foresight: The International Journal of Applied Forecasting* 4: 43–46.

———. 2018. *Mcomp: Data from the M-Competitions. R package version 2.8.* https://CRAN.R-project.org/package=Mcomp.

———. 2020. *fpp2: Data for "Forecasting: Principles and Practice" (2nd Edition). R package version 2.4.* https://CRAN.R-project.org/package=fpp2.

———. 2021. *Forecasting Impact.* Forecasting Impact. https://forecastingimpact.buzzsprout.com/1641538/7708129-rob-hyndman.

———. 2023. *fpp3: Data for "Forecasting: Principles and Practice" (3rd Edition). R package version 0.5.* https://CRAN.R-project.org/package=fpp3.

Hyndman, Rob J., Roman A. Ahmed, George Athanasopoulos, and Han Lin Shang. 2011. "Optimal Combination Forecasts for Hierarchical Time Series." *Computational Statistics & Data Analysis* 55 (9): 2579–89. https://doi.org/10.1016/j.csda.2011.03.006.

Hyndman, Rob J., and George Athanasopoulos. 2014. "Optimally Reconciling Forecasts in a Hierarchy." *Foresight: The International Journal of Applied Forecasting* 35 (42-48).

———. 2021. *Forecasting: Principles and Practice.* 3rd ed. Melbourne, Australia: OTexts. https://otexts.com/fpp3/.

Hyndman, Rob J., George Athanasopoulos, Christoph Bergmeir, Gabriel Caceres, Leanne Chhay, Mitchell O'Hara-Wild, Fotios Petropoulos, Slava Razbash, Earo Wang, and Farah Yasmeen. 2023. *forecast: Forecasting Functions for Time Series and Linear Models. R Package Version 8.21.* https://pkg.robjhyndman.com/forecast/.

Hyndman, Rob J., and Anne B. Koehler. 2006. "Another Look at Measures of Forecast Accuracy." *International Journal of Forecasting* 22 (4): 679–88. https://doi.org/10.1016/j.ijforecast.2006.03.001.

Hyndman, Rob J., Anne B. Koehler, J. Keith Ord, and Ralph D. Snyder. 2008. *Forecasting with Exponential Smoothing: The State Space Approach.* Springer Series in Statistics. New York, NY: Springer. https://doi.org/10.1007/978-3-540-71918-2.

Hyndman, Rob J., and Andrey V. Kostenko. 2007. "Minimum Sample Size Requirements for Seasonal Forecasting Models." *Foresight: The International Journal of Applied Forecasting*, no. 6: 12–15.

Hyndman, Rob J., and Nikolaos Kourentzes. 2018. *thief: Temporal HIErarchical Forecasting. R package version 0.3.* http://pkg.robjhyndman.com/thief.

Hyndman, Rob J., Alan J. Lee, and Earo Wang. 2016. "Fast Computation of Reconciled Forecasts for Hierarchical and Grouped Time Series." *Computational Statistics & Data Analysis* 97: 16–32. https://doi.org/10.1016/j.csda.2015.11.007.

Hyndman, Rob J., Alan Lee, Earo Wang, and Shanika Wickramasuriya. 2021. *hts: Hierarchical and Grouped Time Series. R package version 6.0.2.* https://CRAN.R-project.org/package=hts.

Hyndman, Rob J., and Yangzhuoran Yang. 2023. *tsdl: Time Series Data Library. R package version 0.1.0.* https://github.com/FinYang/tsdl.

Ibrahim, Rouba, Han Ye, Pierre L'Ecuyer, and Haipeng Shen. 2016. "Modeling and Forecasting Call Center Arrivals: A Literature Survey and a Case Study." *International Journal of Forecasting* 32 (3): 865–74. https://doi.org/10.1016/j.ijforecast.2015.11.012.

James, Gareth, Daniela Witten, Trevor Hastie, and Robert Tibshirani. 2021. *An Introduction to Statistical Learning.* 2nd ed. Vol. 112. Springer. https://hastie.su.domains/ISLR2/ISLRv2_website.pdf.

Januschowski, Tim, Jan Gasthaus, Yuyang Wang, David Salinas, Valentin Flunkert, Michael Bohlke-Schneider, and Laurent Callot. 2020. "Criteria for Classifying Forecasting Methods (Invited Commentary on the M4 Forecasting Competition)." *International Journal of Forecasting* 36 (1): 167–77. https://doi.org/10.1016/j.ijforecast.2019.05.008.

Januschowski, Tim, Stephan Kolassa, Martin Lorenz, and Christian Schwarz. 2013. "Forecasting with in-Memory Technology." *Foresight: The International Journal of Applied Forecasting* 31: 14–20.

Januschowski, Tim, Yuyang Wang, Kari Torkkola, Timo Erkkilä, Hilaf Hasson, and Jan Gasthaus. 2022. "Forecasting with Trees." *International Journal of Forecasting* 38 (4): 1473–81. https://doi.org/10.1016/j.ijforecast.2021.10.004.

Kaggle. 2023. "Advertising Sales Dataset." Kaggle website. https://www.kaggle.com/datasets/yasserh/advertising-sales-dataset.

Kahneman, Daniel. 2012. *Thinking: Fast and Slow.* Penguin.

Kahneman, Daniel, Dan Lovallo, and Olivier Sibony. 2011. "Before You Make That Big Decision." *Harvard Business Review* 89 (6): 51–60.

Kang, Yanfei, Rob J. Hyndman, and Kate Smith-Miles. 2017. "Visualising Forecasting Algorithm Performance Using Time Series Instance Spaces." *International Journal of Forecasting* 33 (2): 345–58. https://doi.org/10.1016/j.ijforecast.2016.09.004.

Kolassa, Stephan. 2008. "Can We Obtain Valid Benchmarks from Published Surveys of Forecast Accuracy?" *Foresight: The International Journal of Applied Forecasting*, no. 11: 6–14. https://ideas.repec.org/a/for/ijafaa/y2008i11p6-14.html.

———. 2014. "Data Science Without Knowledge of a Specific Topic, Is It Worth Pursuing as a Career?" DataScience.StackExchange. https://datascience.stackexchange.com/a/2406/.

———. 2016a. "Evaluating Predictive Count Data Distributions in Retail Sales Forecasting." *International Journal of Forecasting* 32 (3): 788–803. https://doi.org/10.1016/j.ijforecast.2015.12.004.

———. 2016b. "Sometimes It's Better to Be Simple Than Correct." *Foresight: The International Journal of Applied Forecasting* 40: 20–26.

———. 2017. "What Are the Shortcomings of the Mean Absolute Percentage Error (MAPE)?" Cross Validated. https://stats.stackexchange.com/q/299712.

———. 2020. "Will Deep and Machine Learning Solve Our Forecasting Problems?" *Foresight: The International Journal of Applied Forecasting* 57: 13–18.

———. 2022a. "Commentary on the M5 forecasting competition." *International Journal of Forecasting* 38 (4): 1562–68. https://doi.org/10.1016/j.ijforecast.2021.08.006.

———. 2022b. "Selecting ARIMA Orders by ACF/PACF Vs. By Information Criteria." Cross Validated. https://stats.stackexchange.com/q/595150.

———. 2022c. "Do We Want Coherent Hierarchical Forecasts, or Minimal MAPEs or MAEs? (We Won't Get Both!)." *International Journal of Forecasting*, December. https://doi.org/10.1016/j.ijforecast.2022.11.006.

———. 2023a. "How We Deal with Zero Actuals Has a Huge Impact on the MAPE and Optimal Forecasts." *Foresight: The International Journal of Applied Forecasting* 69: 13–16.

———. 2023b. "Minitutorial: The Pinball Loss for Quantile Forecasts." *Foresight: The International Journal of Applied Forecasting*, no. 68: 66–67.

Kolassa, Stephan, and Rob J. Hyndman. 2010. "Free Open-Source Forecasting Using R." *Foresight: The International Journal of Applied Forecasting* 17: 19–24.

Kolassa, Stephan, and Roland Martin. 2011. "Percentage Errors Can Ruin Your Day (and Rolling the Dice Shows How)." *Foresight: The International Journal of Applied Forecasting* 23: 21–29. https://ideas.repec.org/a/for/ijafaa/y2011i23p21-27.html.

Kolassa, Stephan, and Wolfgang Schütz. 2007. "Advantages of the MAD/Mean Ratio over the MAPE." *Foresight: The International Journal of Applied Forecasting* 6: 40–43. https://ideas.repec.org/a/for/ijafaa/y2007i6p40-43.html.

Kolassa, Stephan, and Enno Siemsen. 2016. *Demand Forecasting for Managers.* Supply and Operations Management Collection. New York, NY: Business Expert Press. https://www.businessexpertpress.com/books/demand-forecasting-managers/.

Kourentzes, Nikolaos, Devon Barrow, and Fotios Petropoulos. 2019. "Another Look at Forecast Selection and Combination: Evidence from Forecast Pooling." *International Journal of Production Economics* 209: 226–35. https://doi.org/10.1016/j.ijpe.2018.05.019.

Kourentzes, Nikolaos, and Fotios Petropoulos. 2016. "Forecasting with Multivariate Temporal Aggregation: The Case of Promotional Modelling." *International Journal of Production Economics* 181, Part A: 145–53. https://doi.org/10.1016/j.ijpe.2015.09.011.

———. 2022. *MAPA: Multiple Aggregation Prediction Algorithm. R package version 2.0.5.* https://CRAN.R-project.org/package=MAPA.

Kourentzes, Nikolaos, Fotios Petropoulos, and Juan R. Trapero. 2014. "Improving Forecasting by Estimating Time Series Structural Components Across Multiple Frequencies." *International Journal of Forecasting* 30 (2): 291–302. https://doi.org/10.1016/j.ijforecast.2013.09.006.

Kremer, Mirko, Brent Moritz, and Enno Siemsen. 2011. "Demand Forecasting Behavior: System Neglect and Change Detection." *Management Science* 57 (10): 1827–43. https://doi.org/10.1287/mnsc.1110.1382.

Kremer, Mirko, Enno Siemsen, and Douglas J. Thomas. 2016. "The Sum and Its Parts: Judgmental Hierarchical Forecasting." *Management Science* 62 (9): 2745–64. https://doi.org/10.1287/mnsc.2015.2259.

Kreye, M. E., Y. M. Goh, L. B. Newnes, and Paul Goodwin. 2012. "Approaches to Displaying Information to Assist Decisions Under Uncertainty." *Omega* 40 (6): 682–92. https://doi.org/10.1016/j.omega.2011.05.010.

Lapide, Larry. 2014. "S&OP : The Process Revisited." *Journal of Business Forecasting* 34 (3): 12–16.

Larrick, Richard P., and Jack B. Soll. 2006. "Intuitions about Combining Opinions: Misappreciation of the Averaging Principle." *Management Science* 52 (1): 111–27. https://doi.org/10.1287/mnsc.1050.0459.

Lawrence, Michael, Paul Goodwin, and Robert Fildes. 2002. "Influence of User Participation on DSS Use and Decision Accuracy." *Omega* 30 (5): 381–92. https://doi.org/10.1016/s0305-0483(02)00048-8.

Lawrence, Michael, and Spyros Makridakis. 1989. "Factors Affecting Judgmental Forecasts and Confidence Intervals." *Organizational Behavior and Human Decision Processes* 43 (2): 172–87. https://doi.org/10.1016/0749-5978(89)90049-6.

Lawrence, Michael, Marcus O'Connor, and Bob Edmundson. 2000. "A Field Study of Sales Forecasting Accuracy and Processes." *European Journal of Operational Research* 122: 151–60. https://doi.org/10.1016/S0377-2217(99)00085-5.

Löning, Markus, Anthony Bagnall, Sajaysurya Ganesh, Viktor Kazakov, Jason Lines, and Franz J Király. 2019. "Sktime: A Unified Interface for Machine Learning with Time Series." *arXiv Preprint arXiv:1909.07872.*

Makridakis, Spyros. 1993. "Accuracy Measures: Theoretical and Practical Concerns." *International Journal of Forecasting* 9 (4): 527–29. https://doi.org/10.1016/0169-2070(93)90079-3.

Makridakis, Spyros, Chris Chatfield, Michèle Hibon, Michael Lawrence, Terence Mills, Keith Ord, and LeRoy F. Simmons. 1993. "The M2-Competition: A Real-Time Judgmentally Based Forecasting Study." *International Journal of Forecasting* 9 (1): 5–23. https://doi.org/10.1016/0169-2070(93)90044-N.

Makridakis, Spyros, and Michèle Hibon. 2000. "The M3-Competition: Results, Conclusions and Implications." *International Journal of Forecasting* 16 (4): 451–76. https://doi.org/10.1016/S0169-2070(00)00057-1.

Makridakis, Spyros, Fotios Petropoulos, and Evangelos Spiliotis. 2022. "The M5 Competition: Conclusions." *International Journal of Forecasting*, May. https://doi.org/10.1016/j.ijforecast.2022.04.006.

Makridakis, Spyros, Evangelos Spiliotis, and Vassilios Assimakopoulos. 2022. "M5 Accuracy Competition: Results, Findings, and Conclusions." *International Journal of Forecasting* 38 (4): 1346–64. https://doi.org/10.1016/j.ijforecast.2021.11.013.

Makridakis, Spyros, Evangelos Spiliotis, Vassilios Assimakopoulos, Zhi Chen, Anil Gaba, Ilia Tsetlin, and Robert L. Winkler. 2022. "The M5 uncertainty competition: Results, findings and conclusions." *International Journal of Forecasting* 38 (4): 1365–85. https://doi.org/10.1016/j.ijforecast.2021.10.009.

Mannes, Albert E., and Don A. Moore. 2013. "A Behavioral Demonstration of Overconfidence in Judgment." *Psychological Science* 24 (7): 1190–97. https://doi.org/10.1177/0956797612470700.

Matejka, Justin, and George Fitzmaurice. 2017. "Same Stats, Different Graphs: Generating Datasets with Varied Appearance and Identical Statistics Through Simulated Annealing." In *Proceedings of the 2017 CHI Conference on Human Factors in Computing Systems*, 1290–94. https://www.autodesk.com/research/publications/same-stats-different-graphs.

Matthews, Robert. 2000. "Storks Deliver Babies ($p = 0.008$)." *Teaching Statistics* 22 (2): 36–38. https://doi.org/10.1111/1467-9639.00013.

McCarthy, Teresa M., Donna F. Davis, Susan L. Golicic, and John T. Mentzer. 2006. "The Evolution of Sales Forecasting Management: A 20-Year Longitudinal Study of Forecasting Practices." *Journal of Forecasting* 25 (5): 303–24. https://doi.org/10.1002/for.989.

Mélard, Guy. 2014. "On the Accuracy of Statistical Procedures in Microsoft Excel 2010." *Computational Statistics* 29 (5): 1095–1128. https://doi.org/10.1007/s00180-014-0482-5.

Mello, John. 2009. "The Impact of Sales Forecast Game Playing on Supply Chains." *Foresight: The International Journal of Applied Forecasting* 13: 13–22.

Miller, Don, and Dan Williams. 2003. "Shrinkage Estimators of Time Series Seasonal Factors and Their Effect on Forecasting Accuracy." *International Journal of Forecasting* 19: 669–84. https://doi.org/10.1016/S0169-2070(02)00077-8.

Mohammadi, Hamid. 2022. *croston: croston model for intermittent time series. Python package version 0.1.2.4.* https://pypi.org/project/croston/.

Mohammadipour, Maryam, John E. Boylan, and Aris A. Syntetos. 2012. "The Application of Product-Group Seasonal Indexes to Individual Products." *Foresight: The International Journal of Applied Forecasting* 26: 18–24.

Moritz, Brent, Enno Siemsen, and Mirko Kremer. 2014. "Judgmental Forecasting: Cognitive Reflection and Decision Speed." *Production and Operations Management* 23 (7): 1146–60. https://doi.org/10.1111/poms.12105.

Morlidge, Steve. 2014a. "Do Forecasting Methods Reduce Avoidable Error? Evidence from Forecasting Competitions." *Foresight: The International Journal of Applied Forecasting* 32: 34–39.

———. 2014b. "Forecast Quality in the Supply Chain." *Foresight: The International Journal of Applied Forecasting* 33: 26–31.

———. 2015. "Measuring the Quality of Intermittent Demand Forecasts: It's Worse Than We've Thought!" *Foresight: The International Journal of Applied Forecasting* 37: 37–42.

Nahmias, Steven. 1994. "Demand Estimation in Lost Sales Inventory Systems." *Naval Research Logistics (NRL)* 41 (6): 739–57. https://doi.org/10.1002/1520-6750(199410)41:6%3C739::AID-NAV3220410605%3E3.0.CO;2-A.

Nahmias, Steven, and Tava Lennon Olsen. 2015. *Production and Operations Analysis.* 7th ed. Waveland Press.

Nikolopoulos, Konstantinos, Aris A. Syntetos, John E. Boylan, Fotios Petropoulos, and Vassilis Assimakopoulos. 2011. "An Aggregate–Disaggregate Intermittent Demand Approach (ADIDA) to Forecasting: An Empirical Proposition and Analysis." *Journal of the Operational Research Society* 62 (3): 544–54. https://doi.org/10.1057/jors.2010.32.

O'Hara-Wild, Mitchell, Rob J. Hyndman, and Earo Wang. 2022. *feasts: Feature Extraction and Statistics for Time Series. R package version 0.3.0.* https://CRAN.R-project.org/package=feasts.

O'Hara-Wild, Mitchell, Rob J. Hyndman, Earo Wang, and Gabriel Caceres. 2020. *fable: Forecasting Models for Tidy Time Series. R Package Version 0.2.1.* https://fable.tidyverts.org/.

Oliva, Rogelio, and Noel Watson. 2009. "Managing Functional Biases in Organizational Forecasts: A Case Study of Consensus Forecasting in Supply Chain Planning." *Production and Operations Management* 18 (2): 138–51. https://doi.org/10.1111/j.1937-5956.2009.01003.x.

Önkal, Dilek, Paul Goodwin, Mary Thomson, Sinan Gönül, and Andrew Pollock. 2009. "The Relative Influence of Advice from Human Experts and Statistical Methods on Forecast Adjustments." *Journal of Behavioral Decision Making* 22 (4): 390–409. https://doi.org/10.1002/bdm.637.

Ord, Keith, Robert Fildes, and Nikolaos Kourentzes. 2017. *Principles of Business Forecasting.* 2nd ed. Wessex Press.

Panagiotelis, Anastasios, Puwasala Gamakumara, George Athanasopoulos, and Rob J. Hyndman. 2023. "Probabilistic Forecast Reconciliation: Properties, Evaluation and Score Optimisation." *European Journal of Operational Research* 306 (2): 693–706. https://doi.org/10.1016/j.ejor.2022.07.040.

Pearl, Judea, and Dana Mackenzie. 2018. *The Book of Why: The New Science of Cause and Effect.* Basic Books.

Pedregosa, F., G. Varoquaux, A. Gramfort, V. Michel, B. Thirion, O. Grisel, M. Blondel, et al. 2011. "Scikit-Learn: Machine Learning in Python." *Journal of Machine Learning Research* 12: 2825–30.

Perktold, Josef, Skipper Seabold, Jonathan Taylor, and statsmodels-developers. 2022. *Statsmodels: Statistical Models, Hypothesis Tests, and Data Exploration. Python Package Version 0.13.5.* https://www.statsmodels.org/stable/index.html#.

Petropoulos, Fotios, Daniele Apiletti, Vassilios Assimakopoulos, M. Zied Babai, Devon K Barrow, Souhaib Ben Taieb, Christoph Bergmeir, et al. 2022. "Forecasting: Theory and Practice." *International Journal of Forecasting* 38 (3): 705–871. https://doi.org/10.1016/j.ijforecast.2021.11.001.

Petropoulos, Fotios, Yael Grushka-Cockayne, Enno Siemsen, and Evangelos Spiliotis. 2022. "Wielding Occam's Razor: Fast and Frugal Retail Forecasting." Working Paper. https://doi.org/10.48550/ARXIV.2102.13209.

Qin, Yan, Ruoxuan Wang, Asoo J. Vakharia, Yuwen Chen, and Michelle M. H. Seref. 2011. "The Newsvendor Problem: Review and Directions for Future Research." *European Journal of Operational Research* 213 (2): 361–74. https://doi.org/10.1016/j.ejor.2010.11.024.

R Core Team. 2022. *R: A Language and Environment for Statistical Computing.* Vienna, Austria: R Foundation for Statistical Computing. https://www.R-project.org/.

Raiffa, Howard. 1968. *Decision Analysis.* Reading, MA: Addison-Wesley.

Richardson, Ronny. 2011. *Business Applications of Multiple Regression.* New York, NY: Business Expert Press.

Rickwalder, Dan. 2006. "Forecasting Weekly Effects of Recurring Irregular Occurrences." *Foresight: The International Journal of Applied Forecasting* 4: 16–18.

Robette, Johann. 2023. "Does Improved Forecast Accuracy Translate to Business Value?" *Foresight: The International Journal of Applied Forecasting* 68: 12–19.

Robinson, Lynn A. 2006. *Trust Your Gut: How the Power of Intuition Can Grow Your Business.* Chicago, IL: Kaplan Publishing.

Rostami-Tabar, Bahman, M. Zied Babai, and Aris A. Syntetos. 2022. "To Aggregate or Not to Aggregate: Forecasting of Finite Autocorrelated Demand." *Journal of the Operational Research Society.* https://doi.org/10.1080/01605682.2022.2118631.

Salinas, David, Valentin Flunkert, Jan Gasthaus, and Tim Januschowski. 2020. "DeepAR: Probabilistic Forecasting with Autoregressive Recurrent Networks." *International Journal of Forecasting* 36 (3): 1181–91. https://doi.org/10.1016/j.ijforecast.2019.07.001.

Satchell, Stephen E., and Soosung Hwang. 2016. "Tracking Error: Ex Ante Versus Ex Post Measures." Edited by Stephen Satchell, 54–62. https://doi.org/10.1007/978-3-319-30794-7_4.

Schaer, Oliver, Ivan Svetunkov, Alisa Yusupova, and Robert Fildes. 2022. "Survey: Forecasting Software Trends in a Challenging World." Institute for Operations Research; the Management Sciences (INFORMS). https://doi.org/10.1287/orms.2022.05.17.

Schapire, Robert E. 1990. "The Strength of Weak Learnability." *Machine Learning* 5 (2): 197–227. https://doi.org/10.1007/bf00116037.

Schauberger, Philipp, and Alexander Walker. 2022. *Openxlsx: Read, Write and Edit Xlsx Files. R Package Version 4.2.5.1.* https://CRAN.R-project.org/package=openxlsx.

Scheele, Lisa M., Ulrich W. Thonemann, and Marco Slikker. 2017. "Designing Incentive Systems for Truthful Forecast Information Sharing Within a Firm." *Management Science* 64 (8): 3690–3713. https://doi.org/10.1287/mnsc.2017.2805.

Schmidt, Torsten, and Simeon Vosen. 2013. "Forecasting Consumer Purchases Using Google Trends." *Foresight: The International Journal of Applied Forecasting* 30: 38–41.

Schubert, Sean. 2012. "Forecastability: A New Method for Benchmarking and Driving Improvement." *Foresight: The International Journal of Applied Forecasting* 26: 5–13.

scikit-learn core developers. 2023. *scikit-learn: Machine Learning in Python. Python package version 1.2.1.* https://scikit-learn.org/stable/index.html.

Seaman, Brian, and John Bowman. 2022. "Applicability of the M5 to Forecasting at Walmart." *International Journal of Forecasting* 38 (4): 1468–72. https://doi.org/10.1016/j.ijforecast.2021.06.002.

Seifert, Matthias, Enno Siemsen, Allègre L. Hadida, and Andreas B. Eisingerich. 2015. "Effective Judgmental Forecasting in the Context of Fashion Products." *Journal of Operations Management* 36: 33–45. https://doi.org/10.1016/j.jom.2015.02.001.

Shmueli, Galit. 2016. *Practical Time Series Forecasting: A Hands-on Guide.* 3rd ed. Axelrod Schnall.

Shmueli, Galit, and Kenneth C. Lichtendahl. 2018. *Practical Time Series Forecasting with R: A Hands-on Guide.* 2nd ed. Axelrod Schnall.

Silver, Edward A., David F. Pyke, and Douglas J. Thomas. 2017. *Inventory and Production Management in Supply Chains.* 4th ed. CRC Press.

Singh, Sujit. 2013. "Supply Chain Forecasting & Planning: Move on from Microsoft Excel?" *Foresight: The International Journal of Applied Forecasting* 31: 6–13.

Slowikowski, Kamil. 2023. *Ggrepel: Automatically Position Non-Overlapping Text Labels with 'Ggplot2'. R Package Version 0.9.3.* https://CRAN.R-project.org/package=ggrepel.

Smith, Joe. 2009. "The Alignment of People, Process, and Tools." *Foresight: The International Journal of Applied Forecasting* 15: 13–18.

Smith, Taylor G. et al. 2022. *pmdarima: ARIMA estimators for Python. Python package version 2.0.2.* http://alkaline-ml.com/pmdarima/.

Soyer, Emre, and Robin M. Hogarth. 2012. "The Illusion of Predictability: How Regression Statistics Mislead Experts." *International Journal of Forecasting* 28 (3): 695–711. https://doi.org/10.1016/j.ijforecast.2012.02.002.

Spiegel, Alica. 2014. "So You Think You're Smarter Than a CIA Agent." http://www.npr.org/blogs/parallels/2014/04/02/297839429/-so-you-think-youre-smarter-than-a-cia-agent.

Spiliotis, Evangelos. 2022. "Decision Trees for Time-Series Forecasting." *Foresight: The International Journal of Applied Forecasting* 64: 30–44.

Stoffer, David, and Nicky Poison. 2023. *astsa: Applied Statistical Time Series Analysis. R package version 2.0.* https://CRAN.R-project.org/package=astsa.

Surowiecki, James. 2004. *The Wisdom of Crowds.* New York, NY: Anchor.

Svetunkov, Ivan. 2023. *smooth: Forecasting Using State Space Models. R package version 3.2.0.* https://CRAN.R-project.org/package=smooth.

Svetunkov, Ivan, and John E. Boylan. 2020. "State-Space ARIMA for Supply-Chain Forecasting." *International Journal of Production Research* 58 (3): 818–27. https://doi.org/10.1080/00207543.2019.1600764.

Syntetos, Aris A., M. Zied Babai, John E. Boylan, Stephan Kolassa, and Konstantinos Nikolopoulos. 2016. "Supply Chain Forecasting: Theory, Practice, Their Gap and the Future." *European Journal of Operational Research* 252 (1): 1–26. https://doi.org/10.1016/j.ejor.2015.11.010.

Syntetos, Aris A., M. Zied Babai, and Everette Shaw Gardner Jr. 2015. "Forecasting Intermittent Inventory Demands: Simple Parametric Methods Vs. Bootstrapping." *Journal of Business Research* 68 (8): 1746–52. https://doi.org/10.1016/j.jbusres.2015.03.034.

Syntetos, Aris A., M. Zied Babai, David Lengu, and Nezih Altay. 2011. "Distributional Assumptions for Parametric Forecasting of Intermittent Demand." In *Service Parts Management*, edited by Nezih Altay and Lewis A. Litteral, 31–52. Springer London. https://doi.org/10.1007/978-0-85729-039-7_2.

Syntetos, Aris A., and John E. Boylan. 2001. "On the Bias of Intermittent Demand Estimates." *International Journal of Production Economics* 71: 457–66. https://doi.org/10.1016/S0925-5273(00)00143-2.

———. 2005. "The Accuracy of Intermittent Demand Estimates." *International Journal of Forecasting* 21 (2): 303–14. https://doi.org/10.1016/j.ijforecast.2004.10.001.

Syntetos, Aris A., Konstantinos Nikolopoulos, and John E. Boylan. 2010. "Judging the Judges Through Accuracy-Implication Metrics: The Case of Inventory Forecasting." *International Journal of Forecasting* 26 (1): 134–43.

Taleb, Nassim Nicholas. 2014. *Antifragile: Things That Gain from Disorder.* Vol. 3. Random House Trade Paperbacks.

Taylor, James W. 2003. "Short-Term Electricity Demand Forecasting Using Double Seasonal Exponential Smoothing." *Journal of the Operational Research Society* 54 (8): 799–805. https://doi.org/10.1057/palgrave.jors.2601589.

———. 2010. "Exponentially Weighted Methods for Forecasting Intraday Time Series with Multiple Seasonal Cycles." *International Journal of Forecasting* 26 (4): 627–46. https://doi.org/10.1016/j.ijforecast.2010.02.009.

Taylor, James W., and Ralph D. Snyder. 2012. "Forecasting Intraday Time Series with Multiple Seasonal Cycles Using Parsimonious Seasonal Exponential Smoothing." *Omega* 40 (6): 748–57. https://doi.org/10.1016/j.omega.2010.03.004.

Taylor, P. F., and M. E. Thomas. 1982. "Short Term Forecasting: Horses for Courses." *Journal of the Operational Research Society* 33 (8): 685–94. https://doi.org/10.1057/jors.1982.157.

Taylor, Sean J., and Benjamin Letham. 2018. "Forecasting at Scale." *The American Statistician* 72 (1): 37–45. https://doi.org/10.1080/00031305.2017.1380080.

Taylor, Sean, and Ben Letham. 2021. *Prophet: Automatic Forecasting Procedure. R Package Version 1.0.* https://CRAN.R-project.org/package=prophet.

Tetlock, Philip E., and Dan Gardner. 2015. *Superforecasting*. Crown Publishers.

Teunter, Ruud, and Babangida Sani. 2009. "On the Bias of Croston's Forecasting Method." *European Journal of Operational Research* 194 (1): 177–83. https://doi.org/10.1016/j.ejor.2007.12.001.

The LaTeX community. n.d. *The LaTeX Project*. https://www.latex-project.org/.

Tim. 2017. "How to Know That Your Machine Learning Problem Is Hopeless?" Cross Validated. https://stats.stackexchange.com/q/222179.

Timme, Stephen G., and Christine Williams-Timme. 2003. "The Real Cost of Holding Inventory." *Supply Chain Management Review* 7 (4): 30–37. http://www.scmr.com/article/CA318372.html.

Tonetti, Bill. 2006. "Tips for Forecasting Semi-New Products." *Foresight: The International Journal of Applied Forecasting* 4: 54–56.

US Census Bureau. n.d. *X-13ARIMA-SEATS*. https://www.census.gov/data/software/x13as.html.

Vandeput, Nicolas. 2023. *Demand Forecasting: Best Practices*. Manning Publications.

Wang, Xiaoqian, Rob J. Hyndman, Feng Li, and Yanfei Kang. 2022. "Forecast Combinations: An over 50-Year Review," May. https://arxiv.org/abs/2205.04216.

Wickham, Hadley, Mara Averick, Jennifer Bryan, Winston Chang, Lucy McGowan, Romain François, Garrett Grolemund, et al. 2019. "Welcome to the Tidyverse." *Journal of Open Source Software* 4 (43): 1686. https://doi.org/10.21105/joss.01686.

Xie, Yihui. 2022. *bookdown: Authoring Books and Technical Documents with R Markdown. R package version 0.30.* https://github.com/rstudio/bookdown.

Yardley, Elizabeth, and Fotios Petropoulos. 2021. "Beyond Error Measures to the Utility and Cost of the Forecasts." *Foresight: The International Journal of Applied Forecasting*, no. 63: 36–45.

Zhu, Hao. 2021. *kableExtra: Construct Complex Table with 'Kable' and Pipe Syntax. R Package Version 1.3.4.* https://CRAN.R-project.org/package=kableExtra.

Index

Printed in the United States
by Baker & Taylor Publisher Services

Printed in the United States
by Baker & Taylor Publisher Services